# GCSE
# mathematics

**HIGHER**

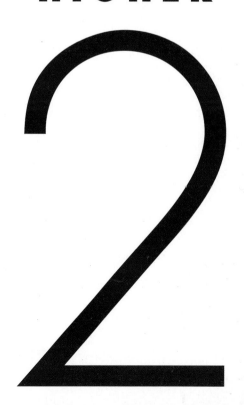

**R. C. Solomon**

## Acknowledgements

The authors and publishers would like to thank all the teachers, schools and advisers who evaluated *CCEA GCSE Mathematics Higher 2* and whose comments contributed so much to this final version.

Particular thanks go to: David Bullock, Frodsham High School, Warrington; Philip Chaffé, King Edward VI School, Lichfield; John D. Collins, Education Consultant and Inspector of Schools; Patrick Gallagher, Convent of Jesus and Mary RC High School, London; Peter Marks, Ilfracombe College, Devon; Kevin Pankhurst, Pilton Community College, Barnstaple; Kim O'Driscoll-Tole, University of Strathclyde.

With special thanks to Alan Davison, Chair of Examiners, GCSE Mathematics, CCEA, Belfast.

## Photo acknowledgements

**Cover:** Christopher Hill Photographic; **p.30** English Heritage Photographic Library © Skyscan Balloon Photography; **p.46** © British Museum, London; **p.177** AKG London; **p.180** Science Photo Library; **p.186** AKG London; **p.218** Mary Evans Picture Library.

© R. C. Solomon/MATHS NOW! Writing Group 2002

First published in 2002
by John Murray (Publishers) Ltd
50 Albemarle Street
London W1S 4BD

All rights reserved. No part of this publication may be reproduced in any material form (including photocopying or storing in any medium by electronic means and whether or not transiently or incidentally to some other use of this publication) without the written permission of the publisher, except in accordance with the provisions of the Copyright, Designs and Patents Act 1988 or under the terms of a licence issued by the Copyright Licensing Agency.

Layouts by Stephen Rowling/springworks
Artwork by Oxford Designers & Illustrators

Typeset in 10/12 Times by Wearset Ltd, Boldon, Tyne and Wear
Printed and bound in Spain by Book Print S.L., Barcelona

A catalogue entry for this title is available from the British Library

**ISBN 0 7195 7674 1**
Teacher's Resource Book GCSE Higher 2 0 7195 7675 X

# Contents

| | |
|---|---|
| Acknowledgements | ii |
| Foreword | vi |
| Introduction | vii |

## 1 Algebraic methods — 1

| | |
|---|---|
| Expanding and factorising | 1 |
| Factorising quadratics | 2 |
| Powers | 3 |
| Algebraic fractions | 4 |
| Expressions with four terms | 5 |

## 2 Trigonometry: the sine and cosine rules — 8

| | |
|---|---|
| Right-angled triangles | 8 |
| The sine rule | 14 |
| The cosine rule | 19 |
| Solving triangles | 23 |
| The sine and cosine rules in three dimensions | 25 |

## 3 Quadratics — 32

| | |
|---|---|
| Solving quadratic equations by factorising | 32 |
| Completing the square | 34 |
| Solving equations by completing the square | 38 |
| The quadratic formula | 40 |
| Practical uses | 42 |

## 4 Probability — 48

| | |
|---|---|
| Combinations of outcomes | 48 |
| Tree diagrams | 51 |
| Drawing with and without replacement | 57 |

## 5 Circles — 63

| | |
|---|---|
| Basic circle theorems | 63 |
| Angle at the centre | 68 |
| Angles on the same arc | 71 |
| Cyclic quadrilaterals | 74 |
| Alternate segment theorem | 76 |
| Mixed examples | 79 |

## Proof — 86
- Terminology — 86
- Geometric proof — 88
- Algebraic proof — 96

## 6 Statistical diagrams — 98
- Histograms — 99
- Box plots — 105
- Stem-and-leaf diagrams — 110

## 7 Coordinates — 117
- Gradient and intercepts — 117
- Parallel and perpendicular lines — 120
- Distance between points — 123
- Simultaneous linear and quadratic equations — 125
- Circles — 127

## 8 Transformations of graphs — 132
- Translating graphs — 132
- Stretching graphs — 140
- Reflections — 144
- Combined transformations — 146

## 9 Trigonometry for all angles — 152
- The unit circle — 154
- Finding angles — 157
- Negative angles — 160
- Angles greater than 360° — 162
- Transformations of trigonometric graphs — 163

## 10 Sequences — 169
- Linear sequences — 170
- Exponential sequences — 173
- Quadratic sequences — 175

## 11 Sampling — 182
- Samples — 183
- Time series — 187

## 12 Vectors — 193
- Vectors as translations — 193
- Arithmetic and geometry of vectors — 194
- Vectors and coordinate geometry — 198
- Vectors in pure geometry — 201

Contents v

### 13 Rational and irrational numbers — 208

Rational numbers — 208
Irrational numbers — 210
Arithmetic of surds — 212
Use of surds — 214

## Making and spending money — 219

Making money — 219
Spending money — 221
Taxes — 224

## Review section — 229

### Review module 1 – Number — 230

Fractions — 230
Factors and multiples — 232
Error and limits of accuracy — 234
Approximation — 236
Appropriate accuracy — 237
Standard form — 239
Proportion — 242

### Review module 2 – Algebra — 245

Equations in one unknown — 245
Simultaneous equations — 248
Trial and improvement — 250
Changing the subject — 252
Inequalities — 254
Linear programming — 258
Graphs — 260

### Review module 3 – Shape and space — 266

Lines and angles — 266
Constructions and loci — 272
Pythagoras' theorem — 277
Length, area and volume — 282
Dimensions — 289
Transformations — 290
Similarity — 294
Three-dimensional geometry — 297

### Review module 4 – Handling data — 300

Probability — 300
Scatter diagrams — 303
Cumulative frequency — 306

## Index — 310

# Foreword

In recent years, there have been important changes in the form and content of secondary school mathematics, together with changing ideas about how the subject should be taught. At GCSE level, for example, both teachers and students now need to use a range of information and communication technologies, notably calculators and computers, and to become competent working with spreadsheets and databases.

The author of this book has extensive experience of teaching and examining mathematics at GCSE level. In preparing this book he has worked closely with Alan Davison, CCEA Chair of Examiners in GCSE Mathematics, and with the publishers to ensure that it completely and thoroughly matches our specification for 2003 and beyond, and meets the needs of all students preparing for this new examination.

No one can pretend that mathematics is an easy subject but in the hands of a skilful teacher a good book can make the subject come alive for students and help raise the standard of their mathematical attainment.

I am grateful to all those involved in the preparation of this book and I commend it most warmly to all those concerned with teaching and learning GCSE Mathematics in Northern Ireland.

**Gavin Boyd**
**Chief Executive CCEA**
**June 2001**

# Introduction

This is the second year of your GCSE course. By the end of the year you will be taking the GCSE exam. This book contains all the material necessary to prepare you for that exam. But Mathematics, or any other subject, is not just about passing exams. This book contains much material to show the connection of Mathematics with other areas of life.

Some of you will receive no more formal Mathematics education after this year. Others will continue to study the subject at A level and possibly beyond. In either case, we hope that the Mathematics you have studied so far has given you an insight into the richness and power of the subject.

## Symbols

The symbols used in the Student's Book are as follows:

 Use a calculator

 Use a graphics calculator

 Do not use a calculator

 This is a particularly challenging question/exercise

 Ma1

# 1 Algebraic methods

Throughout this book you will need to be confident and accurate in the use of algebra, as virtually all parts of mathematics involve it in one form or another. This introductory chapter revises the algebraic techniques necessary for the rest of the book.

## Expanding and factorising

We need to **expand** or **factorise** algebraic expressions in order to solve equations, and for many other purposes.

### Expanding

In an expression like $a(b + c)$, the $a$ must multiply both $b$ and $c$.

$$a(b + c) = ab + ac$$

In an expression like $(x + y)(a + b)$, both terms in the first brackets must multiply both terms in the second brackets.

$$(x + y)(a + b) = xa + xb + ya + yb$$

---

*Example*    Expand:    **a** $3x(2x + 4)$    **b** $(2x + 3)(3x - 4)$

**a** Multiply both $2x$ and $4$ by $3x$. Recall that $x \times x$ is $x^2$.

$$3x(2x + 4) = 3x \times 2x + 3x \times 4$$
$$= 6x^2 + 12x$$

$3x(2x + 4) = 6x^2 + 12x$.

**b** Both terms in the first brackets multiply both terms in the second brackets. Before simplification, there are four terms in all.

$$(2x + 3)(3x - 4) = 2x \times 3x - 2x \times 4 + 3 \times 3x - 3 \times 4$$
$$= 6x^2 - 8x + 9x - 12$$
$$= 6x^2 + x - 12$$

$(2x + 3)(3x - 4) = 6x^2 + x - 12$.

---

## Exercise 1.1

Expand the following, simplifying where possible.

1. $2(x - y)$
2. $3(a + b)$
3. $x(a - b)$
4. $p(x + y)$
5. $2(3a + 2b)$
6. $5(2x - 4y)$
7. $2x(3p + 2q)$
8. $3a(3x + 2y)$
9. $2(x + y) + 3(x - y)$
10. $5(a + 2b) + 3(2a - 3b)$
11. $3(2p + 5q) - 2(3p - 5q)$
12. $4(2m - 3n) - 3(5m + 2n)$
13. $x(x + 3)$
14. $a(2 - a)$
15. $y(y + 1)$
16. $k(1 - k)$
17. $(x + m)(y - n)$
18. $(a - b)(c - d)$
19. $(2x - 3y)(4x + 5y)$
20. $(5n + 2m)(3n - 2m)$
21. $(x + 3)(x - 7)$
22. $(2a + 3)(3a + 5)$
23. $(4y - 3)(2y + 5)$
24. $(4k - 5)(2k - 1)$
25. $(3 - 2x)(2 - 3x)$
26. $(5 - 2y)(y - 3)$
27. $(7 - x)(3x - 2)$

# 2 ALGEBRAIC METHODS

## Factorising

Factorisation is the reverse procedure to expansion. Factorise by any factor, numerical or algebraic, that is common to all the terms.

*Example* Factorise: **a** $x^2 - x$ **b** $15x^2y + 6xy^2 + 9xy$

**a** The common factor of both terms is $x$. When we factorise $x$ by itself, the result is 1 (not 0).
$x^2 - x = x(x - 1)$.

**b** The highest common numerical factor is 3. The highest common algebraic factor is $xy$.
$15x^2y + 6xy^2 + 9xy = 3xy(5x + 2y + 3)$.

### Exercise 1.2

Factorise the following.

1. $a^2 + ab$
2. $mn - n^2$
3. $a^2 + 5a$
4. $2x - x^2$
5. $a^2 + a$
6. $y - 3y^2$
7. $x^2 + x^3$
8. $6x^2 + 9xy$
9. $15ab + 21a^2$
10. $8mn^2 - 10m^2n$
11. $\pi r^2 + \pi rs$
12. $4\pi r^2 + 2\pi rh$
13. $5a^3 + 25a^2 + 125a$
14. $7x^3y - 21xy^3$
15. $4pqr + 6p^2q + 10qr^2$
16. $3f^2gh + 9fg^2h - 15fgh^2$
17. $10a^3bc - 5ab^4c + 15abc^5$
18. $3x^3y^2z^2 + 6x^2y^3z^2 - 12x^2y^2z^3$

## Factorising quadratics

An expression of the form $ax^2 + bx + c$, where $a$, $b$ and $c$ are constant, is a **quadratic expression**. An equation of the form $ax^2 + bx + c = 0$ is a **quadratic equation**.

### Method

Take the simpler case of $x^2 + bx + c$, i.e. $a = 1$. Find the factors of $c$. If you can find a pair whose sum is $b$ then the expression factorises.

Note that if $c$ is negative then one factor is negative and the other is positive, so you are looking for a pair of numbers whose *difference* is $b$.

In the general case of $ax^2 + bx + c$, find the factors of $ac$. If you can find a pair whose sum is $b$ then the expression factorises. Write $bx$ in terms of the sum or difference of these factors.

*Example* Factorise: **a** $x^2 - 11x + 30$ **b** $x^2 + 4x - 12$ **c** $3x^2 + 2x - 8$

**a** Here $c$ is positive. The factors of 30 which have a sum of 11 are 5 and 6. Because the $11x$ term is negative, take $-5$ and $-6$.
$x^2 - 11x + 30 = (x - 5)(x - 6)$.

**b** Here $c$ is negative. The factors of 12 which have a difference of 4 are 2 and 6. The $4x$ term is positive, so take $+6$ and $-2$.
$x^2 + 4x - 12 = (x + 6)(x - 2)$.

**c** The product $ac$ is $-24$. This is negative, so we want factors of 24 whose difference is 2. These are 6 and 4.

$$3x^2 + 2x - 8 = 3x^2 + 6x - 4x - 8$$
$$= 3x(x + 2) - 4(x + 2)$$
$$= (x + 2)(3x - 4)$$

$3x^2 + 2x - 8 = (x + 2)(3x - 4)$.

> After a factorisation, you can always check your answer by expanding it.

> Be careful with negative numbers here.

## Exercise 1.3

Factorise the following.

1. $x^2 - 13x + 30$
2. $x^2 - 9x + 20$
3. $x^2 + 8x + 15$
4. $x^2 + 11x + 24$
5. $x^2 + 3x - 28$
6. $x^2 + 5x - 14$
7. $x^2 - 13x - 30$
8. $x^2 - 4x - 21$
9. $2x^2 + 7x + 3$
10. $3x^2 + 17x + 10$
11. $2x^2 - 13x + 15$
12. $6x^2 - 11x + 4$
13. $2x^2 + 3x - 5$
14. $3x^2 + 2x - 8$
15. $3x^2 - 4x - 15$
16. $4x^2 - 4x - 15$

## Powers

The symbol $a^n$ means $n$ copies of $a$, multiplied together. This definition is extended to negative, zero and fractional values of $n$.

### Definitions and results

$$a^{-n} = \frac{1}{a^n} \qquad a^0 = 1 \ (a \neq 0) \qquad a^{1/n} = \sqrt[n]{a}$$

$$a^m \times a^n = a^{m+n} \qquad a^m \div a^n = a^{m-n} \qquad (a^m)^n = a^{mn} \qquad (ab)^n = a^n b^n$$

Using these results, an expression involving powers of 2, 4, 8, etc. can often be reduced to a simpler expression involving powers of 2 only. For example, $8^n = (2^3)^n = 2^{3n}$.

*Example* Simplify the expression $\dfrac{27^x}{9^y \times 3^z}$.

Note that 3, 9 and 27 are all powers of 3.

$$\frac{27^x}{9^y \times 3^z} = \frac{(3^3)^x}{(3^2)^y \times 3^z}$$

$$= \frac{3^{3x}}{3^{2y} \times 3^z}$$

$$= 3^{3x-2y-z}$$

$$\frac{27^x}{9^y \times 3^z} = 3^{3x-2y-z}.$$

## Exercise 1.4

Simplify the following expressions.

1. $(x^2)^3 \times (x^4)^5$
2. $(y^3)^2 \div (y^4)^3$
3. $(3x^{-1}y^2)^3$
4. $(2x)^3 \times (4x)^2$
5. $8(2x)^{-2}$
6. $(3a^{-2})^{-3}$
7. $x \times \sqrt{x} \times \sqrt[3]{x}$
8. $\sqrt{a^3 \times a^5}$
9. $\sqrt{\dfrac{x^8}{x^3}}$
10. $\sqrt{x \times \sqrt{x}}$
11. $\dfrac{a^3 \times b^2}{a^{-4} \times \sqrt{b}}$
12. $\sqrt{ab} \times \sqrt{\dfrac{a}{b}}$
13. $(x^2 y^3)^{-2} \times (x^3 y^{-4})^3$
14. $\dfrac{(6xy)^2}{(2yz)^4}$
15. $2^x \times 4^y \times 8^z$
16. $\dfrac{2^p \times 16^q}{4^r}$
17. $\sqrt{5^{2x} \times 25^x}$
18. $27^a \times 9^{2a} \times 3^{3a}$

# ALGEBRAIC METHODS

## Algebraic fractions

**Algebraic fractions** obey the same rules of arithmetic as ordinary fractions.

Multiplying: $\dfrac{a}{b} \times \dfrac{c}{d} = \dfrac{ac}{bd}$

Dividing: $\dfrac{a}{b} \div \dfrac{c}{d} = \dfrac{a}{b} \times \dfrac{d}{c} = \dfrac{ad}{bc}$

Adding and subtracting: $\dfrac{a}{b} \pm \dfrac{c}{d} = \dfrac{ad \pm bc}{bd}$

It may be possible to simplify an algebraic fraction, by dividing numerator and denominator by the same term.

In the rule for adding and subtracting, the common denominator is $bd$. There may be a simpler common denominator than this.

*Example* Simplify: **a** $\dfrac{3x}{y} \div \dfrac{9x}{z}$  **b** $\dfrac{3}{x^2} + \dfrac{2}{xy}$

**a** Turn the second fraction upside down and multiply.

$$\dfrac{3x}{y} \div \dfrac{9x}{z} = \dfrac{3x}{y} \times \dfrac{z}{9x} = \dfrac{3xz}{9xy}$$

$3x$ cancels from top and bottom.

$$\dfrac{3x}{y} \div \dfrac{9x}{z} = \dfrac{z}{3y}.$$

**b** The simplest common denominator (the simplest expression both $x^2$ and $xy$ will divide into) is $x^2y$.

$$\dfrac{3}{x^2} + \dfrac{2}{xy} = \dfrac{3y}{x^2y} + \dfrac{2x}{x^2y}$$

$$\dfrac{3}{x^2} + \dfrac{2}{xy} = \dfrac{3y + 2x}{x^2y}.$$

**Note.** If we had chosen $x^3y$ as the common denominator, we would have obtained

$$\dfrac{3xy + 2x^2}{x^3y}$$

$x$ can be cancelled from top and bottom, giving the same answer as before.

## Exercise 1.5

Simplify the following expressions.

1  $\dfrac{3x}{2y} \times \dfrac{6y}{z}$

2  $\dfrac{2ab}{cd} \times \dfrac{ac}{bd}$

3  $\dfrac{2x^2}{3y} \times \dfrac{5y}{4x}$

4  $\dfrac{xy^2}{2} \times \dfrac{4x}{y}$

5  $\dfrac{x+2}{x} \times \dfrac{x-1}{x+2}$

6  $\dfrac{x+1}{x-3} \times \dfrac{x+2}{2x+2}$

# Expressions with four terms

7. $\dfrac{3x-3}{x+1} \times \dfrac{3x+3}{x-1}$
8. $\dfrac{\pi r^2 h}{\frac{4}{3}\pi r^3}$
9. $\dfrac{x^2-1}{(x+1)^2}$
10. $\dfrac{x^2+3x+2}{x^2+5x+4}$
11. $\dfrac{a}{b} \div \dfrac{c}{b}$
12. $\dfrac{3x}{2y} \div \dfrac{9a}{4b}$
13. $\dfrac{x}{y} \div \dfrac{x^2}{y^2}$
14. $\dfrac{\pi r^3}{h} \div \dfrac{\pi r^2 h}{s}$
15. $\dfrac{x}{3} + \dfrac{y}{4}$
16. $\dfrac{a}{5} - \dfrac{b}{2}$
17. $\dfrac{3\pi r}{2} - \dfrac{5\pi r}{6}$
18. $\dfrac{1}{x} + \dfrac{1}{y}$
19. $\dfrac{3}{a} - \dfrac{2}{b}$
20. $\dfrac{1}{xy} + \dfrac{1}{x^2}$
21. $\dfrac{2}{x^2 y} - \dfrac{3}{xy^2}$
22. $\dfrac{a}{x} + \dfrac{b}{xy}$
23. $\dfrac{x}{pq} - \dfrac{y}{pqr}$
24. $\dfrac{x+3}{2} + \dfrac{x+2}{3}$
25. $\dfrac{x-2}{5} - \dfrac{x-1}{3}$
26. $\dfrac{x+1}{2} - \dfrac{x+3}{6}$
27. $\dfrac{1}{x+1} + \dfrac{2}{x+2}$
28. $\dfrac{2}{x} + \dfrac{1}{x+3}$
29. $\dfrac{3}{x-2} - \dfrac{2}{x-1}$
30. $\dfrac{4}{x+2} - \dfrac{3}{x-1}$

## Expressions with four terms

Look at the expansion of $(x+a)(y+b)$.

$$(x+a)(y+b) = xy + xb + ay + ab$$

The result is an expression with four terms, which cannot be simplified by collecting like terms. If we want to *factorise* such an expression, we group the terms in pairs and factorise them separately.

$$xy + xb + ay + ab = x(y+b) + a(y+b)$$
$$= (x+a)(y+b) \quad \text{as there is a common factor of } (y+b)$$

---

*Examples*  Factorise $km + 2n + kn + 2m$.

Rearrange, putting the terms with $k$ together.

$$km + kn + 2m + 2n$$

Factorise the first two terms and the last two terms.

$$k(m+n) + 2(m+n)$$

Now there is a common factor of $(m+n)$. Factorise by it. It is multiplied by $k$ and by 2.

$$km + 2n + kn + 2m = (m+n)(k+2).$$

# ALGEBRAIC METHODS

Factorise $t^2 + 6ab - 2at - 3bt$.

Rearrange as $t^2 - 2at - 3bt + 6ab$.
$t$ is a factor of the first pair, and $3b$ is a factor of the second pair. Be careful with the negative terms. When we divide $+6ab$ by $-3b$, it becomes $-2a$.

$$t(t - 2a) - 3b(t - 2a)$$
$$t^2 + 6ab - 2at - 3bt = (t - 2a)(t - 3b).$$

**Notes**
1 All these factorisations can be checked by expansion. For the example above

$$(t - 2a)(t - 3b) = t \times t + t \times (-3b) + (-2a) \times t + (-2a) \times (-3b)$$
$$= t^2 - 3bt - 2at + 6ab$$

After a rearrangement of terms this is the same as the original expression.

2 If we group the terms in pairs in a different way the final result will be equivalent. For the example above

$$t^2 + 6ab - 2at - 3bt = t^2 - 3bt + 6ab - 2at$$
$$= t(t - 3b) + 2a(3b - t)$$
$$= t(t - 3b) - 2a(t - 3b) \quad \text{to ensure that there is a}$$
$$= (t - 3b)(t - 2a) \quad \text{common factor of } (t - 3b)$$

This is the same as the result above, with the factors in a different order.

## Exercise 1.6

Factorise the expressions in questions 1–16.

1  $ch + ck + dh + dk$
2  $pr + qs + ps + qr$
3  $xn - ym - xm + yn$
4  $ac - bc + bd - ad$
5  $3hu + 2kv + hv + 6ku$
6  $xy + 3yk - 2xh - 6hk$
7  $5mv - 10mn + uv - 2nu$
8  $xy + 6pq - 3py - 2qx$
9  $3hn - 5km + kh - 15mn$
10  $15am - 4xn + 10xm - 6an$
11  $15ah - 10bh + 6bg - 9ag$
12  $x^2 + 6ab + 2ax + 3bx$
13  $4xy + t^2 - 2xt - 2yt$
14  $k^2 - 5wk + 3kv - 15vw$
15  $6yb + p^2 - 2bp - 3yp$
16  $6xa - 35yb - 14xb + 15ya$

17 In all the expressions above there were either no negative terms or two negative terms. Explain why there were no expressions with one or three negative terms.

## SUMMARY

- To remove the brackets in an expression, **expand** it.
- To write an expression as a product of terms, **factorise** it.
- Expressions can sometimes be simplified using the rules of powers.
- **Algebraic fractions** obey the same rules as numerical fractions.

# Exercise 1A

1. Expand the expression $3(3x - 7y)$.
2. Expand the expression $(3a - 2b)(4a - 5b)$.
3. Factorise the expression $4a^2b + 6ab^2$.
4. Factorise the expression $x^2 - 2x - 80$.
5. Simplify the expression $(3xy)^2 \times \sqrt{x^3 y^{-5}}$.
6. Write $4^x \times 2^{3x-1} \div 8^{1-x}$ as a power of 2.
7. Simplify the fraction $\dfrac{a^2 - b^2}{a^2 - 2ab + b^2}$.
8. Write as a single fraction: $\dfrac{3}{2x} + \dfrac{5}{3y}$.
9. Factorise the expression $xy + 8 + 2x + 4y$.
10. Factorise the expression $6ab + 15 - 9b - 10a$.

# Exercise 1B

1. Expand the expression $2a(a - 3a^2)$.
2. Expand the expression $(2x + 3)(3x - 7)$.
3. Factorise the expression $8x^2 + 4x^3 - 2x$.
4. Factorise the expression $3x^2 + 13x - 10$.
5. Simplify the expression $(3x^{-2}y^{-3})^{-1}$.
6. Write $\sqrt{9^a \times 3^{2b-4}}$ as a power of 3.
7. Simplify the fraction $\dfrac{2\pi r^2 + 2\pi rh}{4\pi r^2}$.
8. Write as a single fraction: $\dfrac{4ax}{3by} \times \dfrac{9cy}{2dx}$.
9. Factorise the expression $6x^2 - 3xm - mn + 2xn$.
10. Factorise the expression $3ay - 10xb + 2ab - 15xy$.

# 2 Trigonometry: the sine and cosine rules

The trigonometric functions, **sine**, **cosine** and **tangent**, enable us to find sides and angles in right-angled triangles. In this chapter we will extend their use to all triangles, whether they are right-angled or not.

If we know enough about a triangle, we can construct it. For example, if we know the three sides of a triangle, then we can construct it and measure the angles. If we know the angles of a triangle and one side, then we can construct it and measure the other sides.

But construction is a slow and inaccurate process. The measurement of a length or an angle can only be done to a limited degree of accuracy, say to the nearest millimetre or the nearest degree. It is quicker and more accurate to *calculate* the unknown angles and sides of a triangle. There are two rules which enable us to do this, the sine rule and the cosine rule.

## Right-angled triangles

First we have a revision of trigonometry in right-angled triangles.

Suppose we have a triangle which is right-angled at $B$. Suppose we either know or want to know one of the other angles, $A$. Label the sides of the triangle as follows:

HYP  the longest side of the triangle, $AC$
OPP  the side opposite $A$, $BC$
ADJ  the side adjacent to (next to) $A$, $AB$

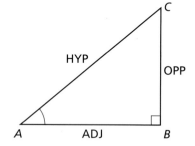

The trigonometric functions are defined as follows:

$$\sin A = \frac{\text{OPP}}{\text{HYP}}$$

$$\cos A = \frac{\text{ADJ}}{\text{HYP}}$$

$$\tan A = \frac{\text{OPP}}{\text{ADJ}}$$

So, if you know a side of a right-angled triangle and one other angle, you can find the other sides. If you know two sides of a right-angled triangle, you can find the angles.

Suppose you are looking up at something. The angle between the horizontal and your line of sight is the **angle of elevation**.

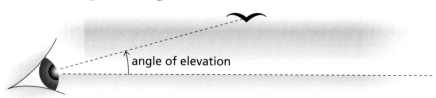

## Right-angled triangles

Suppose you are looking down at something. The angle between the horizontal and your line of sight is the **angle of depression**.

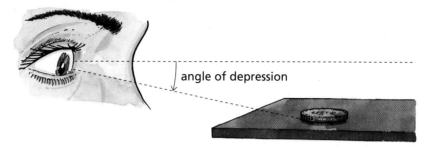

Note that in both cases the angle is measured with the horizontal, not the vertical. Trigonometry is also used in questions involving bearings.

> **Remember:**
> 1. Make sure that your calculator is set to degree mode (not grads or rads). A small D or deg should appear on the display. Make sure you know how to change to degree mode, if the calculator is wrongly set.
> 2. Make sure you know how your calculator operates. In some models, the angle is entered before the function, in others it is entered after.
> 3. Be careful when finding the inverse function of a fraction. To find $\tan^{-1}\frac{1}{3}$, for example, a correct sequence could be either of
>
>      [1] [÷] [3] [=] [shift] [tan]
>   OR   [shift] [tan] [(] [1] [÷] [3] [)] [=]
>
> An answer beginning 18.4 should appear.
> The following sequences are wrong.
>
>      [1] [÷] [3] [shift] [tan] [=]    0.01397... appears. This is $\frac{1}{\tan^{-1}3}$.
>
>      [shift] [tan] [1] [÷] [3] [=]    15 appears. This is $\frac{\tan^{-1}1}{3}$.

*Examples*    In $\triangle XYZ$, $\angle XZY = 90°$, $XY = 47\,\text{cm}$ and $\angle XYZ = 37°$. Find $YZ$.

Label the triangle as shown. We know the hypotenuse and want to know the adjacent, so the function to use is cosine.

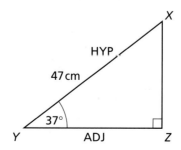

$$\cos 37° = \frac{YZ}{47}$$

$$YZ = \cos 37° \times 47\,\text{cm}$$

$YZ = 37.5\,\text{cm}.$

# 10 TRIGONOMETRY: THE SINE AND COSINE RULES

A ship sails so that it is 37 km south and 42 km west of its starting point. Along what bearing has it been sailing?

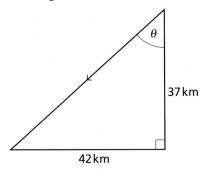

Let $\theta$ be the angle between the direction and south. Then

$$\tan \theta = \tfrac{42}{37}$$

Hence $\theta = \tan^{-1}\left(\tfrac{42}{37}\right) = 48.6°$. Add on $180°$ to find the bearing.
The ship has been sailing on a bearing of $229°$.

---

A telegraph pole leans at an angle of $17°$ to the vertical. The top of the pole is 4.2 m above the ground. How long is the pole?

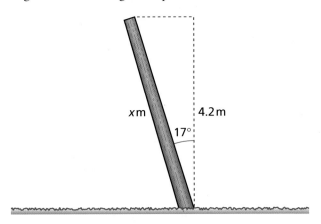

Let the length of the pole be $x$ m. This is the hypotenuse of the triangle, and the height above the ground is the adjacent. Use cosine.

$$\cos 17° = \frac{4.2}{x}$$

Hence $x = 4.2 \div \cos 17°$.

The length of the pole is 4.39 m.

# Right-angled triangles 11

> Tangents to a circle from the same point are equal in length.

$AB$ is a chord of a circle. The tangents at $A$ and $B$ meet at $T$, where $\angle ATB = 34°$ and $AT = 7$ cm. Find the length of $AB$.

$ATB$ is an isosceles triangle.
Let $M$ be the midpoint of $AB$. Then $TMA$ is a right-angled triangle.

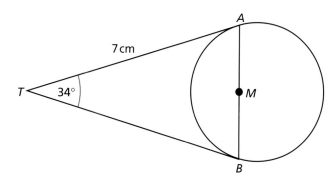

$\angle ATM = \frac{1}{2} \times \angle ATB = 17°$.

$AM = 7 \times \sin 17° = 2.047$ cm

Double to obtain $AB$.
The length of $AB$ is 4.09 cm.

# Exercise 2.1

Mainly revision

**1** Find the sides labelled with letters in these triangles.

**a**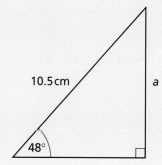

**b**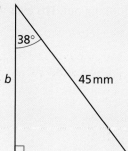

**c**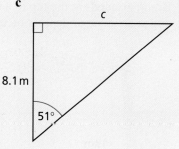

## 12  TRIGONOMETRY: THE SINE AND COSINE RULES

**2** Find the angles labelled with letters in these triangles.

a 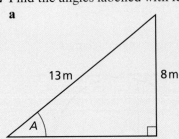  b 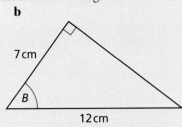  c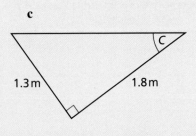

**3** In $\triangle ABC$, $\angle ABC = 90°$, $AC = 20$ cm and $\angle ACB = 48°$. Find $AB$ and $BC$.
**4** In $\triangle LMN$, $\angle MNL = 90°$, $NL = 5$ m and $\angle MLN = 23°$. Find $NM$.
**5** In $\triangle PQR$, $\angle RPQ = 90°$, $RQ = 8$ cm and $PQ = 5$ cm. Find $\angle RQP$.
**6** In $\triangle DEF$, $\angle EFD = 90°$, $EF = 11$ m and $FD = 23$ m. Find $\angle FDE$.
**7** In $\triangle ABC$, $\angle ACB = 90°$, $\angle CBA = 38°$ and $CB = 12$ m. Find $AB$.
**8** In $\triangle LMN$, $\angle MLN = 90°$, $\angle LNM = 42°$ and $LM = 0.32$ m. Find $MN$.
**9** A plank of length 3.2 m leans against a wall. If the angle between the plank and the wall is 31° find how far up the wall the plank reaches.

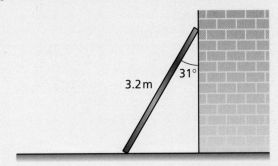

**10** The plank of question 9 now slips 0.2 m down the wall. Find the new angle between the plank and the wall.
**11** Brendan stands at the top of a building looking out. His eyes are 13 m above the ground. He sees a dog at an angle of depression of 17°.

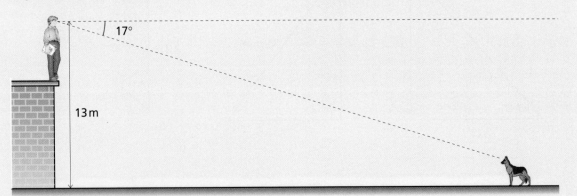

**a** How far away is the dog?
**b** The dog now walks 10 m towards the building. What is the new angle of depression?
**12** Two towers, A and B, are 350 m apart. From the top of A, the angle of elevation of the top of B is 12°. From the top of B, the angle of depression of the bottom of A is 33°. Find the height of A.

## Right-angled triangles

**13** A tower is on a hilltop. From point A, 200 m away from the tower horizontally, the angle of elevation of the top of the tower is 22° and the angle of elevation of the base of the tower is 13°. Find the height of the tower.

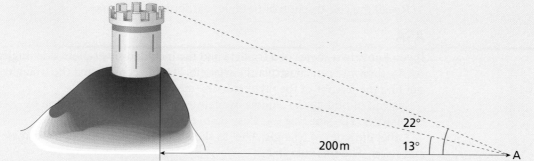

**14** The diagram shows a chord which subtends 32° at the centre of the circle. If the length of the chord is 18 cm, find the radius of the circle.

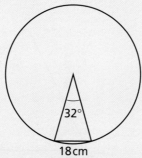

**15** The diagram shows an equilateral triangle inscribed in a circle. The side of the triangle is 16 cm.

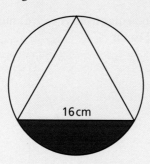

  **a** Find the radius of the circle.
  **b** Find the area of the shaded segment.

**16** A tree trunk is a cylinder of radius 30 cm. It is placed horizontally, and the top 2 cm is planed off. Find the width of the cut surface.

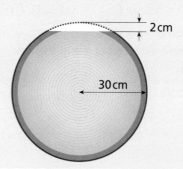

# 14 TRIGONOMETRY: THE SINE AND COSINE RULES

## The sine rule

Not all triangles are right-angled, of course. But you can still use trigonometric functions to find their sides and angles.

Recall the construction of triangles earlier in your course.

### ASA

If you know two angles of a triangle and the included side (angle-side-angle, or ASA), then you can construct the triangle. Having constructed the triangle, you can find the lengths of the other sides by measurement.

Instead of finding the lengths by measurement, you can find them by calculation. The **sine rule** enables you to do this.

Label the sides of triangle $ABC$ as shown, so that side $a$ is opposite angle $A$ and so on. The sine rule states

$$\frac{a}{\sin A} = \frac{b}{\sin B} = \frac{c}{\sin C}$$

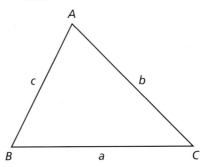

**Proof**

Draw a perpendicular from $A$ to $BC$, meeting it at $D$. Then by the trigonometry you have used so far

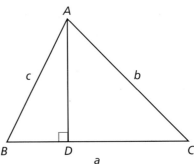

$AD = c \times \sin B$     (in $\triangle ABD$)
$AD = b \times \sin C$     (in $\triangle ACD$)

Hence $c \times \sin B = b \times \sin C$.

Divide both sides by $\sin B$ and by $\sin C$

$$\frac{b}{\sin B} = \frac{c}{\sin C}$$

> We can divide by $\sin B$ and $\sin C$, as neither is 0.

# The sine rule

The other part of the sine rule follows similarly. Try for yourself in the first question of the next exercise.

## Exercise 2.2

1. Complete the proof of the sine rule, by drawing a perpendicular from $B$ to $CA$, or from $C$ to $AB$.
2. Here is another proof of the sine rule. Recall that the area of a triangle can be written as $\frac{1}{2}ab \sin C$, or as $\frac{1}{2}bc \sin A$, or as $\frac{1}{2}ca \sin B$. Put two of these equal to each other, then rearrange.

Now we show how to use the sine rule.

### Using the sine rule to find unknown sides

*Example* In triangle $ABC$, $\angle C = 47°$, $\angle B = 76°$ and $AC = 7.8$ cm. Find $BC$.

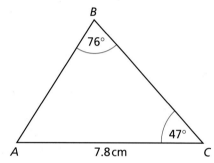

The third angle is given by
$$\angle A = 180° - 47° - 76° = 57°.$$

Now use the part of the rule that we need. We know all three angles and $AC$, which is $b$. We want to know $BC$, which is opposite angle $A$ and so is written $a$.

$$\frac{a}{\sin 57°} = \frac{7.8}{\sin 76°}$$

Hence
$$a = \frac{7.8 \times \sin 57°}{\sin 76°}$$
$$= 6.741$$

$BC = 6.74$ cm.

## Exercise 2.3

**You will need:**
- ruler
- protractor

1. Draw accurately a triangle $ABC$ with $AB = 6$ cm, $\angle ABC = 50°$, $\angle BAC = 70°$. Find the length of $AC$ by measurement, and by calculation using the sine rule. How close are they?

2 Find the sides labelled with letters in these triangles.

a 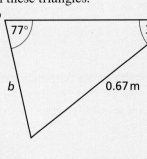 b 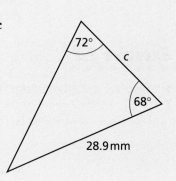 c

3 In $\triangle ABC$, $AB = 8$ m, $\angle ABC = 47°$ and $\angle CAB = 63°$. Find $CA$.
4 In $\triangle DEF$, $FE = 9$ cm, $\angle EFD = 53°$ and $\angle DEF = 72°$. Find $DF$.
5 In $\triangle LMN$, $MN = 62$ mm, $\angle LMN = 81°$ and $\angle MLN = 75°$. Find $LN$.
6 In $\triangle PQR$, $PQ = 18$ cm, $\angle PRQ = 67°$ and $\angle PQR = 61°$. Find $QR$.
7 The diagram shows a stepladder that can be adjusted so that the sides have different lengths. The feet are 1 m apart, and the sides make 78° and 85° with the floor. Find the lengths of the sides.

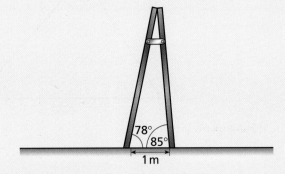

8 Two planks of wood lean against each other as shown. The shorter plank has length 2.5 m and makes an angle of 32° with the ground. The longer plank makes 25° with the ground. Find the length of the longer plank.

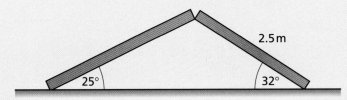

9 I see a church spire on a bearing of 040°. I walk north for 1.5 km, and the spire is now on a bearing of 105°. How far am I now from the spire?

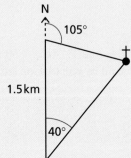

# The sine rule

**10** Ayville is 60 miles west of Beeton. The bearings of Ceebury from Ayville and Beeton are 036° and 344° respectively. Find the distance of Ceebury from Ayville.

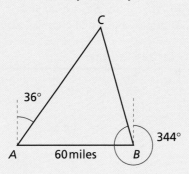

## Using the sine rule to find unknown angles

We have used the sine rule to find unknown sides. It can also be used to find unknown angles. It helps to have the unknown quantities on the numerators of the fractions, so turn the rule upside down, i.e. take the reciprocal of each of the three terms.

$$\frac{\sin A}{a} = \frac{\sin B}{b} = \frac{\sin C}{c}$$

*Example* In $\triangle ABC$, $\angle A = 62°$, $AB = 6$ cm and $BC = 7$ cm. Find $\angle B$.

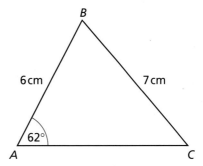

Here we know $AB$, which is $c$, and $BC$, which is $a$. Use part of the sine rule to find $\angle C$.

$$\frac{\sin 62°}{7} = \frac{\sin C}{6}$$

Hence $\sin C = \dfrac{6 \times \sin 62°}{7} = 0.7568$.

Hence $\angle C = \sin^{-1} 0.7568 = 49.18°$.

Subtract from 180° to find $B$.

$$\angle B = 180° - 62° - 49.18°$$

$\angle B = 68.8°$ (to the nearest 0.1°).

**Remember:**
If $\sin x = r$, then $x = \sin^{-1} r$.

# 18 TRIGONOMETRY: THE SINE AND COSINE RULES

## Exercise 2.4

**You will need:**
- ruler
- protractor

1. Draw accurately a triangle $ABC$ with $AB = 5\,\text{cm}$, $BC = 6\,\text{cm}$ and $\angle BAC = 40°$. Find $\angle ABC$ by measurement, and by calculation using the sine rule. Do your results agree?

2. Find the angles labelled with letters in these triangles.

   **a**

   **b**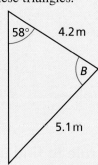

   **c**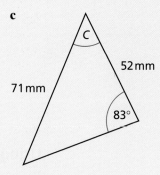

3. In $\triangle ABC$, $AB = 8\,\text{m}$, $BC = 11\,\text{m}$ and $\angle BAC = 52°$. Find $\angle BCA$.
4. In $\triangle DEF$, $EF = 22\,\text{cm}$, $FD = 30\,\text{cm}$ and $\angle DEF = 62°$. Find $\angle FDE$.
5. In $\triangle LMN$, $NL = 66\,\text{mm}$, $ML = 54\,\text{mm}$ and $\angle LMN = 37°$. Find $\angle NLM$.
6. In $\triangle PQR$, $PQ = 14\,\text{cm}$, $PR = 10\,\text{cm}$ and $\angle PRQ = 50°$. Find $\angle RPQ$.

7. The diagram shows a stepladder that has been adjusted so that the lengths of the sides are 2.5 m and 3 m. The shorter side makes 73° with the horizontal. Find the angle that the longer side makes with the horizontal.

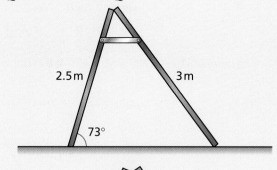

8. Two planks lean against each other as shown. The left plank is 4 m long, and the feet of the planks are 4.5 m apart. The angle between the planks is 58°. Find the angles the planks make with the ground.

9. Fleur stands 400 m west of Charlie. They walk in the directions shown. When they meet, Fleur has walked 300 m, and the angle between their paths is 80°. Find the bearings along which they have walked.

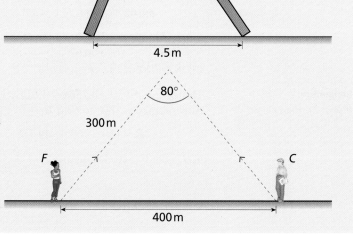

# The cosine rule

Recall again the construction of triangles.

## SAS

If you know two sides of a triangle and the angle enclosed (side-angle-side, or SAS), then you can construct the triangle. Having constructed the triangle, you can find the third side by measurement.

Instead of measuring the length, you can find it by calculation. The **cosine rule** enables you to do this.

With the same convention that side $a$ is opposite angle $A$ and so on, the cosine rule states

$$a^2 = b^2 + c^2 - 2bc\cos A$$

**Proof**

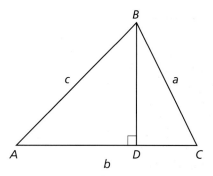

Draw a perpendicular from $B$ to $AC$, meeting it at $D$. Then

$$AD = c \times \cos A \qquad \text{(in } \triangle DAB\text{)}$$
$$BD^2 = c^2 - (c\cos A)^2 \qquad \text{(Pythagoras' theorem in } \triangle DAB\text{)}$$
$$a^2 = CD^2 + BD^2 \qquad \text{(Pythagoras' theorem in } \triangle BCD\text{)}$$

But $CD = b - AD = b - c\cos A$. Substituting for $CD$ and for $BD^2$ in the third equation above, we get

$$a^2 = (b - c\cos A)^2 + (c^2 - (c\cos A)^2)$$
$$a^2 = b^2 - 2bc\cos A + (c\cos A)^2 + c^2 - (c\cos A)^2$$

Cancelling $(c\cos A)^2$ and rearranging we obtain the result

$$a^2 = b^2 + c^2 - 2bc\cos A$$

**Note.** This formula gives $a^2$ in terms of the other sides. The cosine rule could just as well have $b^2$ or $c^2$ as the subject of the formula. These three formulae are all versions of the cosine rule

$$a^2 = b^2 + c^2 - 2bc\cos A \quad b^2 = c^2 + a^2 - 2ca\cos B \quad c^2 = a^2 + b^2 - 2ab\cos C$$

## Using the cosine rule to find unknown sides

*Example*  In triangle $PQR$, $PQ = 5.3$ m, $PR = 4.7$ m and $\angle QPR = 73°$. Find $QR$.

Be careful to follow the convention that each side is labelled with the letter of the opposite angle: write $PQ$ as $r$, $PR$ as $q$ and $QR$ as $p$. Use the cosine rule, with $p^2$ as subject.

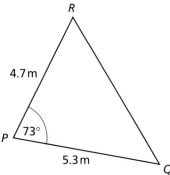

$$p^2 = q^2 + r^2 - 2qr\cos P$$
$$= 4.7^2 + 5.3^2 - 2 \times 4.7 \times 5.3 \times \cos 73°$$
$$= 35.61$$

Hence $p = \sqrt{35.61} = 5.968$.
$QR = 5.97$ m  (to 3 significant figures).

## Exercise 2.5

You will need:
- ruler
- protractor

1. Draw accurately a triangle $ABC$ with $AB = 6$ cm, $AC = 5$ cm and $\angle BAC = 55°$. Find $BC$ by measurement, and by calculation using the cosine rule. Do your results agree?
2. Find the sides labelled with letters in these triangles.

**a**

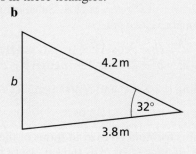

**b**

**c**

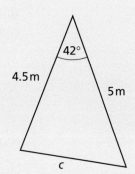

3. In $\triangle ABC$, $AB = 10$ cm, $BC = 15$ cm and $\angle ABC = 70°$. Find $AC$.
4. In $\triangle DEF$, $EF = 9$ m, $ED = 8$ m and $\angle DEF = 80°$. Find $DF$.
5. In $\triangle LMN$, $MN = 43$ mm, $LN = 48$ mm and $\angle LNM = 57°$. Find $LM$.
6. In $\triangle PQR$, $PR = 5.5$ cm, $PQ = 4.7$ cm and $\angle RPQ = 68°$. Find $RQ$.
7. A stepladder is adjusted so that its sides have lengths 2.8 m and 3.2 m. The angle between the sides is 32°. Find the distance between the feet of the sides.
8. Two ships leave port. One sails for 100 km on a bearing of 032°, the other for 80 km on a bearing of 340°. Find the final distance between the ships.

## The cosine rule

**9** From a church door C, Ayton is 8 miles away on a bearing of 172°, and Beeville is 7 miles away on a bearing of 205°. Find the distance between Ayton and Beeville.

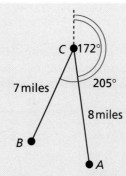

**10** Looking out to sea, a coastguard can see one ship 18 km away on a bearing of 277°, and another 20 km away on a bearing of 352°. What is the distance between the ships?

### Using the cosine rule to find angles

Recall another method of constructing triangles.

#### SSS

If you know the three sides of a triangle (side-side-side, or SSS), you can construct it. Having constructed the triangle, you can measure the angles with a protractor. Instead of measuring the angles, you can calculate them.

In the formula $a^2 = b^2 + c^2 - 2bc\cos A$, make $\cos A$ the subject.

$$2bc\cos A = b^2 + c^2 - a^2$$

$$\cos A = \frac{b^2 + c^2 - a^2}{2bc}$$

This version of the formula gives $\cos A$ in terms of the sides. Similar versions give $\cos B$ and $\cos C$. All these are versions of the cosine rule.

$$\cos A = \frac{b^2 + c^2 - a^2}{2bc} \qquad \cos B = \frac{c^2 + a^2 - b^2}{2ca} \qquad \cos C = \frac{a^2 + b^2 - c^2}{2ab}$$

*Example* In triangle $LMN$, $LM = 8$ cm, $MN = 10$ cm and $NL = 11$ cm. Find $\angle N$.

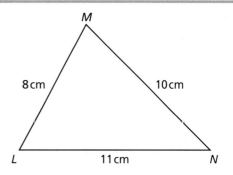

Write $LM$ as $n$, $MN$ as $l$ and $NL$ as $m$. Use the rule, with $\cos N$ as the subject.

$$\cos N = \frac{m^2 + l^2 - n^2}{2ml}$$

$$= \frac{11^2 + 10^2 - 8^2}{2 \times 11 \times 10}$$

$$= \tfrac{157}{220}$$

Hence $N = \cos^{-1}\left(\tfrac{157}{220}\right) = 44.47°$.
$\angle N = 44.5°$ (to the nearest 0.1°).

**Remember:**
If $\cos x = r$, then $x = \cos^{-1} r$.

# 22 TRIGONOMETRY: THE SINE AND COSINE RULES

## Exercise 2.6

You will need:
- ruler
- protractor

1. Draw accurately a triangle $ABC$ with $AB = 5$ cm, $BC = 6$ cm and $CA = 7$ cm. Find $\angle ABC$ by measurement and by calculation. Do your answers agree?
2. Find the angles labelled with letters in these triangles.

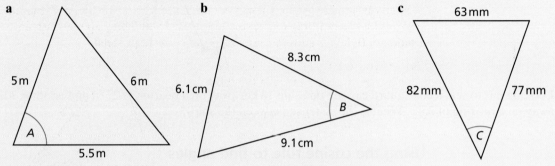

3. In $\triangle ABC$, $AB = 12$ cm, $BC = 15$ cm and $AC = 17$ cm. Find $\angle ABC$.
4. In $\triangle DEF$, $DE = 8$ m, $EF = 10$ m and $FD = 7$ m. Find $\angle DEF$.
5. In $\angle LMN$, $MN = 56$ mm, $ML = 83$ mm and $LN = 77$ mm. Find $\angle MNL$.
6. In $\triangle PQR$, $PR = 22$ cm, $QR = 18$ cm and $PQ = 27$ cm. Find $\angle RQP$.
7. Find the largest angle in a triangle with sides 25 m, 28 m and 32 m. (The largest angle is opposite the largest side.)
8. Find the smallest angle in a triangle with sides 0.3 m, 0.27 m and 0.31 m.
9. A stepladder is adjusted so that its sides have length 3.2 m and 3.5 m. The feet are 1.2 m apart. Find the angle between the sides.
10. Ecksville ($X$) is 50 km west of Wyetown ($Y$). Zeebury ($Z$) is to the north of both towns, and its distances from Ecksville and Wyetown are 60 km and 48 km respectively. Find the bearing of Zeebury from Ecksville.

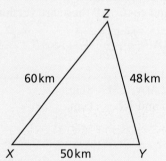

11. Ship A is 200 km north of ship B. Ship C is to the east of both ships, at distances from A and B of 250 km and 320 km respectively. Find the bearing of C from A.

# Solving triangles

There are six numbers to know about a triangle: the lengths of the three sides and the three angles. If you are given some of these numbers you may be able to find the others. This process is **solving** the triangle. For example, if you are given the three sides you can use the cosine rule to find the angles.

In all the situations so far you have been told which rule to use. Often the hardest thing is deciding which rule to use, and which way up the formula should go. Below is a list to help you decide how to proceed.

**SSS** If you know the three sides of a triangle, use the cosine rule in the form
$$\cos A = \frac{b^2 + c^2 - a^2}{2bc}$$
to find the angles.

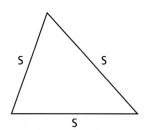

**SAS** If you know two sides of a triangle and the enclosed angle, use the cosine rule in the form $a^2 = b^2 + c^2 - 2bc \cos A$ to find the third side.

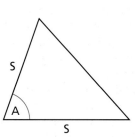

**ASA** If you know two angles of a triangle and one side, use the sine rule in the form
$$\frac{a}{\sin A} = \frac{b}{\sin B} = \frac{c}{\sin C}$$
to find the other sides.

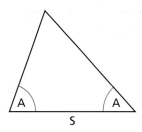

**SSA** If you know two sides of a triangle and an angle not included, use the sine rule in the form
$$\frac{\sin A}{a} = \frac{\sin B}{b} = \frac{\sin C}{c}$$
to find the other angles.

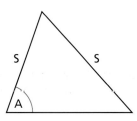

**Notes**
1 Three of these, SSS, SAS and ASA are conditions for congruence: if two triangles have the same sides, for example, then they are congruent.
   The fourth, SSA, is *not* a condition for congruence. See exercise 2E at the end of this chapter.
2 Another tip to help you use the correct rule is this: if you know a side and the opposite angle, use the sine rule. Otherwise use the cosine rule.

Sometimes it is necessary to use the rules more than once.

# 24 TRIGONOMETRY: THE SINE AND COSINE RULES

*Example* In triangle $ABC$, $AB = 6$ cm, $AC = 7$ cm and $\angle A = 58°$. Find $\angle B$.

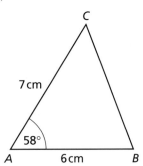

Here we know two sides and the included angle of the triangle (SAS). Use the cosine rule to find the third side, $BC$.

$$BC^2 = 6^2 + 7^2 - 2 \times 6 \times 7 \times \cos 58°$$
$$= 40.487$$

Hence $BC = 6.363$ cm.

Now we know the sides and an angle. Use the sine rule to find $\angle B$.

$$\frac{\sin B}{7} = \frac{\sin 58°}{6.363}$$

$$\sin B = 0.9330$$

$\angle B = 68.9°$ (to the nearest 0.1).

## Exercise 2.7

**1** Find the lengths and angles marked with letters in these triangles.

a. Triangle with sides 14 cm, 13 cm, 11 cm; angle $a$ opposite the 13 cm side (between 14 cm and 11 cm).

b. Triangle with angle 63°, side 57 mm, angle 59°; find side $b$.

c. Triangle with sides 8 m, 10 m, included angle 64°; find side $c$.

d. Triangle with sides 12 m, 11 m, angle 71°; find angle $d$.

**2** In $\triangle ABC$, $AB = 11$ cm, $BC = 16$ cm and $\angle ABC = 43°$. Find $AC$.
**3** In $\triangle DEF$, $DE = 41$ m, $EF = 48$ m and $\angle EDF = 38°$. Find $\angle DEF$.
**4** In $\triangle LMN$, $LM = 12$ cm, $\angle LMN = 63°$ and $\angle LNM = 50°$. Find $LN$.
**5** In $\triangle PQR$, $PQ = 38$ mm, $QR = 32$ mm and $RP = 43$ mm. Find $\angle RPQ$.

6 Find the unknown lengths and angles in the triangles shown.

a 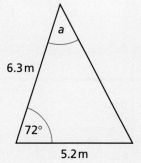  b 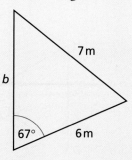  c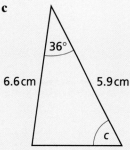

7 In $\triangle ABC$, $AB = 40$ m, $BC = 35$ m and $\angle ABC = 63°$. Find $\angle ACB$.
8 In $\triangle DEF$, $DE = 0.8$ km, $EF = 1.1$ km and $\angle EDF = 42°$. Find $DF$.
9 In $\triangle LMN$, $MN = 47$ mm, $NL = 32$ mm and $\angle NLM = 38°$. Find $ML$.
10 In $\triangle PQR$, $PR = 2.7$ m, $PQ = 3.9$ m and $\angle RPQ = 77°$. Find $\angle RQP$.

# The sine and cosine rules in three dimensions

The sine and cosine rules can be used to find lengths and angles in three-dimensional situations. In all cases use a two-dimensional triangular section.

*Examples*  A tower 100 m high stands in a flat plane. From the top of the tower, village A is on a bearing of 083°, and the angle of depression is 5°. Village B is on a bearing of 118°, and the angle of depression is 4°. Find the distance between the villages.

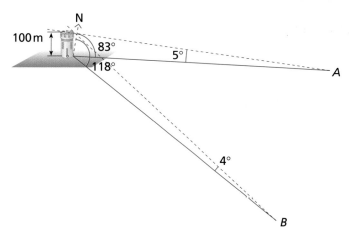

Let the foot of the tower be C. Then $\angle ACB$ is the difference between the bearings, 35°. The distances from C to the villages are

$$AC = 100 \div \tan 5° = 1143 \text{ m} \qquad BC = 100 \div \tan 4° = 1430 \text{ m}.$$

We know two sides of $\triangle ABC$ and the included angle. Use the cosine rule to find the third side.

$$AB^2 = 1143^2 + 1430^2 - 2 \times 1143 \times 1430 \times \cos 35° = 673\,600$$

The distance between the villages is 821 m.

# 26 TRIGONOMETRY: THE SINE AND COSINE RULES

$ABCDEFGH$ is a cuboid with $AB = 4\,\text{cm}$, $AD = 5\,\text{cm}$ and $AE = 6\,\text{cm}$. Find $\angle DEB$.

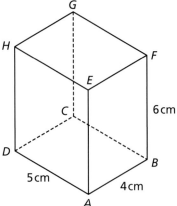

Find $DE$, $EB$ and $BD$ by Pythagoras' theorem.

$$DE = \sqrt{5^2 + 6^2} = \sqrt{61} \qquad EB = \sqrt{4^2 + 6^2} = \sqrt{52}$$
$$BD = \sqrt{4^2 + 5^2} = \sqrt{41}$$

We now know the three sides of the triangle. Use the cosine rule.

$$\cos \angle DEB = \frac{61 + 52 - 41}{2 \times \sqrt{61} \times \sqrt{52}} = 0.639$$

$\angle DEB = 50.3°$.

## Exercise 2.8

1. A cliff is 200 m high. From the top of the cliff two ships are seen: one is at an angle of depression of 3.7° and on a bearing of 352°, the other is at an angle of depression of 2.6° and a bearing of 048°. Find the distance between the ships.
2. A pole of length 5 m with bottom $S$ and top $T$ is held vertical by two ropes $TA$ and $TB$ attached to $T$. $\angle ASB = 120°$. $\angle TAS = 28°$ and $\angle TBS = 35°$. Find
   a  $SA$ and $SB$
   b  $AB$
   c  $TA$ and $TB$
   d  $\angle ATB$

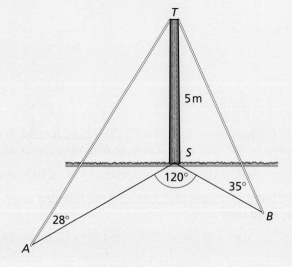

**3** $ABCDEFGH$ is a cuboid, with $AB = 10\,\text{cm}$, $AD = 12\,\text{cm}$ and $AE = 15\,\text{cm}$. Find $\angle EBD$.

**4** $ABCD$ is a tetrahedron. $ABC$ is horizontal, and $D$ is vertically above $A$. $\angle BAC = 60°$, $AC = 7\,\text{cm}$, $AB = 9\,\text{cm}$ and $AD = 12\,\text{cm}$. Find
  **a** $BC$
  **b** $\angle BDC$

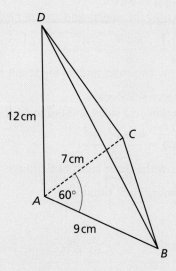

**5** $VABCD$ is a pyramid with a square base $ABCD$ of side $10\,\text{cm}$. $VA = VB = 8\,\text{cm}$, and $VC = VD = 9\,\text{cm}$. The midpoints of $AB$ and $CD$ are $M$ and $N$ respectively. Find
  **a** $VM$
  **b** $VN$
  **c** $\angle VMN$
  **d** the height of the pyramid.

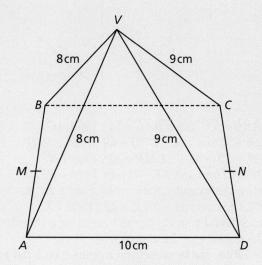

## SUMMARY

- Label a triangle $ABC$ so that side $a$ is opposite angle $A$ and so on.

- The **sine rule** states that $\dfrac{a}{\sin A} = \dfrac{b}{\sin B} = \dfrac{c}{\sin C}$.
  Use this rule to find a side if you know the angles and another side (ASA).

- The sine rule can be written as $\dfrac{\sin A}{a} = \dfrac{\sin B}{b} = \dfrac{\sin C}{c}$.
  Use this to find an angle if you know two sides and an angle not included (SSA).

- The **cosine rule** states that $a^2 = b^2 + c^2 - 2bc \cos A$.
  Use this rule to find a side if you know two sides and the included angle (SAS).

- The cosine rule can be written $\cos A = \dfrac{b^2 + c^2 - a^2}{2bc}$.
  Use this to find an angle if you know the three sides of a triangle (SSS).

- Both rules can be used in three-dimensional situations.

## Exercise 2A

**1** Find the unknown side $x$ in the triangle shown.

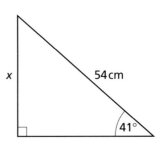

**2** Find the unknown angle $A$ in the triangle shown.

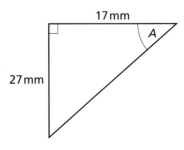

**3** In $\triangle ABC$, $BC = 14$ cm, $\angle ABC = 53°$ and $\angle BAC = 51°$. Find $AC$.

**4** In $\triangle PQR$, $PQ = 43$ mm, $QR = 52$ mm and $\angle RPQ = 48°$. Find $\angle RQP$.

**5** In $\triangle LMN$, $LM = 4.3$ m, $NM = 3.9$ m and $\angle LMN = 83°$. Find $LN$.

**6** In $\triangle DEF$, $DE = 81$ mm, $DF = 67$ mm and $EF = 75$ mm. Find $\angle EFD$.

**7** In $\triangle ABC$, $AB = 11.2$ m, $AC = 12.9$ m and $\angle BAC = 68°$. Find $\angle BCA$.

**8** In $\triangle XYZ$, $YZ = 55$ mm, $ZX = 65$ mm and $\angle XYZ = 43°$. Find $YX$.

**9** Roya stands 800 m north of Sylvia. From Roya, the bearing of a church spire is $098°$, and from Sylvia the bearing is $087°$. Find the distance of the church from Roya.

**10** Two towers, each of height 150 m, can be seen from a point $P$ on a flat plane. The angles of elevation of the tops of the towers from $P$ are $17°$ and $23°$. If the tops of the towers are $T$ and $S$, then $\angle TPS = 39°$. Find the distance between the tops of the towers.

# Exercise 2B

1. Find the unknown angle $A$ in this triangle.

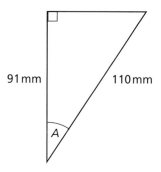

2. Find the unknown side $y$ in this triangle.

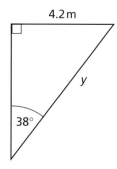

3. In $\triangle IJK$, $JI = 11$ m, $IK = 15$ m and $\angle IJK = 78°$. Find $JK$.
4. In $\triangle ABC$, $BC = 21$ m, $\angle ACB = 61°$ and $\angle CBA = 39°$. Find $AC$.
5. In $\triangle XYZ$, $XY = 0.43$ km, $YZ = 0.41$ km and $ZX = 0.38$ km. Find $\angle YZX$.
6. In $\triangle DEF$, $\angle EFD = 64°$, $FD = 11$ cm and $FE = 13$ cm. Find $ED$.
7. In $\triangle PQR$, $QR = 0.63$ m, $RP = 0.77$ m and $\angle PQR = 53°$. Find $QP$.
8. In $\triangle LMN$, $MN = 5.3$ cm, $NL = 6.8$ cm and $\angle MNL = 81°$. Find $\angle NLM$.
9. A rope 5 m long secures the top of a 2.5 m pole to the ground. The rope is too tight, and has pulled the pole so that it is at 10° to the vertical. Find the angle between the rope and the horizontal.

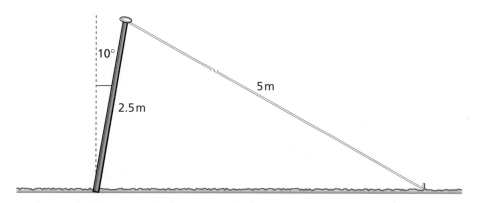

10. $VABCD$ is a pyramid with a square base $ABCD$ of side 2 m. $VA = VB = VC = VD = 3$ m. Let $M$ be the midpoint of $AB$. Find $\angle DVM$.

## TRIGONOMETRY: THE SINE AND COSINE RULES

### Exercise 2C

Suppose a triangle has sides $a$, $b$ and $c$.
Let $s = \tfrac{1}{2}(a+b+c)$. *Hero's formula* for
the area of the triangle is $\sqrt{s(s-a)(s-b)(s-c)}$.

> Hero of Alexandria was a Greek inventor who is best known for his formula for the area of a triangle, and for the invention of the aeolipile, the first steam-powered engine.

1 Suppose a triangle has sides 8 cm, 9 cm, 11 cm. Find its area using Hero's formula. Find one of the angles using the cosine rule, and then find the area of the triangle using the formula $\tfrac{1}{2}ab \sin C$. Do you get the same answer?
2 Repeat question 1 for a triangle with sides 13 m, 16 m and 17 m.

### Exercise 2D

There is much controversy about who first discovered Pythagoras' theorem. Pythagoras himself lived in the sixth century BC, but there are clay tablets from a thousand years earlier which strongly suggest that Babylonian mathematicians were aware of the theorem (though perhaps not of the proof). It has also been suggested that prehistoric people knew of the theorem, and that the layout of the many stone circles of the Megalithic era (such as Stonehenge) provide indications of this.

> **Remember:**
> If a triangle has sides $a$, $b$ and $c$, for which $c^2 = a^2 + b^2$, then the triangle is right-angled.

1 Many of the layouts of stone circles seem to be based on a triangle with sides in the ratio $3:4:5$. Show that this triangle is right-angled.
2 Some of the layouts seem to be based on a triangle with sides in the ratio $12:35:37$. Show that this triangle is right-angled.
3 Some of the layouts seem to be based on a triangle with sides in the ratio $8:9:12$.
   a Is this triangle right-angled?
   b Construct a triangle with sides in this ratio. Does it seem to be right-angled?
   c Use the cosine rule to find the largest angle of the triangle. Is it 90°?
4 Do you think that the prehistoric people who built the stone circles using these layouts were aware of Pythagoras' theorem?

> Use the internet to find out more about Megalithic sites.

# Exercise 2E

In this chapter we have linked the use of the sine or cosine rule with the conditions for construction of a triangle. You may be aware that SSA (two sides and an angle not included) is not always sufficient information to enable us to construct a unique triangle.

You will need:
- ruler
- protractor

1. Construct, as accurately as you can, triangles $ABC$ for which $\angle BAC = 60°$, $AB = 8$ cm and $BC = 7$ cm. Show that there are two possible triangles.

2. Use the sine rule to find $\angle ACB$. For which of the triangles is this the correct value?

3. Measure as accurately as you can the two possible versions of $\angle ACB$. Say they are $\theta$ and $\theta'$. What is the connection between $\theta$ and $\theta'$?

4. If the sine rule works for both triangles, then $\sin \theta$ must equal $\sin \theta'$. Test using a calculator. Test on some other pairs of angles with a similar connection.

*The extension of trigonometric functions to angles greater than 90° is dealt with in more detail in chapter 9.*

# 3 Quadratics

An expression of the form $ax^2 + bx + c$, such as $3x^2 + 2x - 5$, is a quadratic. When we put it equal to 0, obtaining $ax^2 + bx + c = 0$, we have a **quadratic equation**, such as $3x^2 + 2x - 5 = 0$.

- In some cases these equations can be solved exactly. For example, the equation $x^2 - 3x + 2 = 0$ has the exact solutions $x = 1$ and $x = 2$. You have already met this type of equation.
- In other cases the solutions can be found correct to a given degree of accuracy. The equation $x^2 - 2x - 1 = 0$ has solutions $x = 2.414$ and $x = -0.414$, correct to three decimal places. You will meet this sort of equation in this chapter.
- Some quadratic equations cannot be solved by ordinary numbers. The equation $x^2 + x + 2 = 0$ has no real solutions.

## Solving quadratic equations by factorising

*See chapter 1 for how to factorise a quadratic expression.*

If a quadratic expression factorises, then the corresponding equation can be solved. First, ensure that all the terms are on one side of the equation, as $ax^2 + bx + c = 0$. Then factorise the quadratic expression. For example, the quadratic equation $3x^2 + 8x - 16$ is factorised as $(x + 4)(3x - 4)$.

Each of the factors gives a solution: if $(x + 4) = 0$, then $x = -4$; if $(3x - 4) = 0$, then $x = \frac{4}{3}$.

*Examples*

Solve the equation $x^2 + 4x - 12 = 0$.

Factorise the left-hand side:

$$(x + 6)(x - 2) = 0$$

**Remember:**

*When factorising something of the form $x^2 + bx + c$, look for two numbers whose product is c and whose sum is b.*

Either $(x + 6) = 0$, giving $x = -6$, or $(x - 2) = 0$, giving $x = 2$.
$x = -6$ or $x = 2$.

Solve the equation $x + 1 = \dfrac{12}{x}$.

First clear up the fraction, by multiplying by $x$.

$$x^2 + x = 12$$

Take the 12 to the left, then proceed as before.

$$x^2 + x - 12 = 0$$
$$(x + 4)(x - 3) = 0$$

$x = -4$ or $x = 3$.

The length of a rectangle is 5 cm greater than the width. The area is 66 cm². If the width is $x$ cm, find an equation in $x$ and hence find the width.

If the width is $x$ cm, the length is $(x + 5)$ cm. The product of these is the area.

$$x(x + 5) = 66$$

Expand, collect all terms on one side and factorise.

$$x^2 + 5x = 66$$
$$x^2 + 5x - 66 = 0$$
$$(x + 11)(x - 6) = 0$$

So $x = -11$ or $x = 6$. Ignore the negative value.
The width is 6 cm.

## Exercise 3.1

Mainly revision
Solve these equations by factorising.

1. $x^2 + 6x + 5 = 0$
2. $x^2 - 5x + 6 = 0$
3. $x^2 + 2x - 15 = 0$
4. $x^2 - x - 30 = 0$
5. $x^2 + 4x + 4 = 0$
6. $x^2 - 9 = 0$
7. $2x^2 - 5x + 2 = 0$
8. $3x^2 - 2x - 1 = 0$
9. $6x^2 - 5x - 6 = 0$
10. $8x^2 - 2x - 3 = 0$
11. $x^2 + 2x = 35$
12. $2x^2 = 5x - 2$
13. $(x + 3)(x - 2) = 24$
14. $x(x - 7) = 30$
15. $x + 5 = \dfrac{14}{x}$
16. $x + \dfrac{15}{x} = 8$
17. $\dfrac{x + 3}{x - 2} = x - 5$
18. $\dfrac{12}{x} - \dfrac{12}{x + 1} = 1$

19. The length of a rectangle is 2 m greater than the width. The area is 63 m². Find the width.
20. The sides of a right-angled triangle are $x$ cm, $(2x + 2)$ cm and $(3x - 2)$ cm, where the last is the hypotenuse. Find $x$.
21. 20 cm of wire is cut into two pieces, each of which is bent into the perimeter of a square. The total area of the squares is 13 cm². Find the side of the smaller square.
22. Andrew walks a distance of 20 miles at a steady speed of $x$ m.p.h. For the return journey, his speed is 1 m.p.h. less. The total time for the journey is 9 hours. Find $x$.
23. Siobhan drives 120 miles at a steady speed. For the return journey, her speed increases by 10 m.p.h., and the journey takes 1 hour less. Find her speed for the outward journey.

Not every quadratic expression factorises easily, of course. The expression $x^2 - 2x - 1$, for example, cannot be factorised easily. But look at the graph of $y = x^2 - 2x - 1$.

It crosses the $x$ axis at two points, approximately $x = -0.4$ and $x = 2.4$.

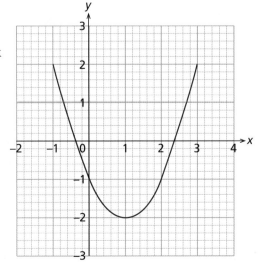

Solving a quadratic equation by a graph is a time consuming and inaccurate method. For example, if your graph paper has 2 cm per unit on the $x$-axis, you cannot solve the equation to better than 1 decimal place. In this chapter we shall find quick and accurate methods of solving such quadratics.

# QUADRATICS

## Completing the square

Consider the expansion of the square of $(x+3)$, i.e. $(x+3)^2$.

$$(x+3)^2 = x^2 + 6x + 9$$

Now consider the expression $x^2 + 6x + 11$. This is 2 greater than the expression above.

$$x^2 + 6x + 11 = x^2 + 6x + 9 + 2 = (x+3)^2 + 2$$

Hence $x^2 + 6x + 11 = (x+3)^2 + 2$. This operation is called **completing the square** of the expression $x^2 + 6x + 11$. The expression $(x+3)^2 + 2$ is the **completed square** form of $x^2 + 6x + 11$.

## Exercise 3.2

1. **a** Write out the expansion of $(x+4)^2$.
   **b** Write $x^2 + 8x + 20$ in terms of this expansion.
   **c** Complete the square of $x^2 + 8x + 20$, i.e. write it in terms of $(x+4)^2$.
2. **a** Write out the expansion of $(x+2)^2$.
   **b** Write $x^2 + 4x + 2$ in terms of this expansion.
   **c** Complete the square of $x^2 + 4x + 2$, i.e. write it in terms of $(x+2)^2$.
3. **a** Write out the expansion of $(x-3)^2$.
   **b** Write $x^2 - 6x + 12$ in terms of this expansion.
   **c** Complete the square of $x^2 - 6x + 12$, i.e. write it in terms of $(x-3)^2$.
4. **a** Write out the expansion of $(x-5)^2$.
   **b** Write $x^2 - 10x + 17$ in terms of this expansion.
   **c** Complete the square of $x^2 - 10x + 17$, i.e. write it in terms of $(x-5)^2$.

In general, suppose we have the expression $x^2 + bx + c$. Consider the expansion of $(x + \tfrac{1}{2}b)^2$.

$$\begin{aligned}(x + \tfrac{1}{2}b)^2 &= x^2 + 2 \times x \times \tfrac{1}{2}b + (\tfrac{1}{2}b)^2 \\ &= x^2 + bx + \tfrac{1}{4}b^2\end{aligned}$$

Compare $x^2 + bx + c$ with $x^2 + bx + \tfrac{1}{4}b^2$. To go from $x^2 + bx + \tfrac{1}{4}b^2$ to $x^2 + bx + c$, add $c$ and take away $\tfrac{1}{4}b^2$. In other words

$$x^2 + bx + c = x^2 + bx + \tfrac{1}{4}b^2 + c - \tfrac{1}{4}b^2$$

Hence $x^2 + bx + c = (x + \tfrac{1}{2}b)^2 + c - \tfrac{1}{4}b^2$.

This formula shows how to complete the square of the expression $x^2 + bx + c$. Take $(x + \tfrac{1}{2}b)^2$, add $c$ and subtract $\tfrac{1}{4}b^2$.

*Example* Complete the square of $x^2 + 8x + 3$.

Here $b = 8$, and $c = 3$. Apply the result above, putting $b = 8$ and $c = 3$. This is the same as considering the expansion of $(x+4)^2$.

$$\begin{aligned}x^2 + 8x + 3 &= (x + \tfrac{1}{2} \times 8)^2 + 3 - \tfrac{1}{4} \times 8^2 \\ &= (x+4)^2 + 3 - 16\end{aligned}$$

$$x^2 + 8x + 3 = (x+4)^2 - 13.$$

## Exercise 3.3

Complete the square for these expressions.

1. $x^2 + 2x + 7$
2. $x^2 + 4x - 1$
3. $x^2 + 6x + 7$
4. $x^2 - 2x + 5$
5. $x^2 - 4x + 1$
6. $x^2 - 6x - 1$
7. $x^2 + 3x + 1$
8. $x^2 + 5x - 2$
9. $x^2 + x + 1$
10. $x^2 - 3x + 5$
11. $x^2 - 5x - 3$
12. $x^2 - x - 1$
13. $x^2 + 2x$
14. $x^2 - 6x$
15. $x^2 + x$
16. $x^2 - 3x$

In all the quadratics considered so far, the coefficient of $x^2$ was 1. If this is not the case, then factorise by the coefficient before completing the square.

*Examples*  Complete the square of $2x^2 + 4x - 3$.

Factorise the first two terms by 2, then complete the square

$$\begin{aligned} 2x^2 + 4x - 3 &= 2(x^2 + 2x) - 3 \\ &= 2((x+1)^2 - 1) - 3 \quad (\text{as } x^2 + 2x = (x+1)^2 - 1) \\ &= 2(x+1)^2 - 2 - 3 \end{aligned}$$

$2x^2 + 4x - 3 = 2(x+1)^2 - 5$.

Complete the square of $3 + 5x - x^2$.

Factorise the last two terms by $-1$, then complete the square.

$$\begin{aligned} 3 + 5x - x^2 &= 3 - (-5x + x^2) \\ &= 3 - (x^2 - 5x) \\ &= 3 - ((x - 2.5)^2 - 6.25) \\ &= 3 - (x - 2.5)^2 + 6.25 \end{aligned}$$

$3 + 5x - x^2 = 9.25 - (x - 2.5)^2$.

## Exercise 3.4

Complete the square for these expressions.

1. $2x^2 + 4x - 1$
2. $2x^2 + 6x + 3$
3. $3x^2 + 6x + 8$
4. $3x^2 - 12x - 5$
5. $5x^2 + 10x + 1$
6. $2x^2 - 2x - 3$
7. $3x^2 + 9x + 8$
8. $2x^2 + x + 2$
9. $2x^2 + 3x - 4$
10. $3x^2 + x + 3$
11. $1 + 2x - x^2$
12. $3 - 2x - x^2$
13. $5 + 4x - x^2$
14. $7 - 6x - x^2$
15. $3 + 4x - 2x^2$
16. $-3 + 8x - 2x^2$
17. $-5 - 6x - 3x^2$
18. $-13 + x - x^2$
19. $9 + x - x^2$
20. $2x - x^2$
21. $-4x - x^2$
22. $6x - 3x^2$
23. $-3x - x^2$

## Maximum and minimum values

Suppose you have completed the square of a quadratic expression. For example, it might be

$$x^2 + 6x - 3 = (x + 3)^2 - 12$$

The square of a number cannot be negative. The expression $(x + 3)^2$ is always at least 0. Hence $(x + 3)^2 - 12$ is always at least $-12$. In symbols:

$$(x + 3)^2 \geq 0. \text{ Hence } (x + 3)^2 - 12 \geq -12$$

$(x + 3)^2 - 12$ reaches the value $-12$ when $(x + 3)^2 = 0$, i.e. when $x = -3$. The least value of $x^2 + 6x - 3$ is $-12$, reached when $x = -3$. This least value is called the **minimum**. So the minimum of $x^2 + 6x - 3$ is $-12$.

The diagram below shows the graph of $y = x^2 + 6x - 3$. Notice that it has its lowest point at $(-3, -12)$. This corresponds to the result found by completing the square. The least value of $x^2 + 6x - 3$ is $-12$, achieved when $x = -3$.

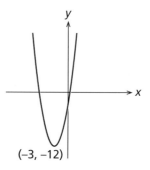

Similarly, consider this expression.

$$5 - 4x - x^2 = 9 - (x + 2)^2$$

The expression $(x + 2)^2$ is always at least 0. So $9 - (x + 2)^2$ is always at *most* 9.

$$(x + 2)^2 \geq 0. \text{ Hence } 9 - (x + 2)^2 \leq 9$$

The greatest value of $5 - 4x - x^2$ is 9, reached when $x = -2$. This greatest value is called the **maximum**. So the maximum of $5 - 4x - x^2$ is 9.

The diagram below shows the graph of $y = 5 - 4x - x^2$. Notice that it has its highest point at $(-2, 9)$. This corresponds to the result found by completing the square. The greatest value of $5 - 4x - x^2$ is 9, achieved when $x = -2$.

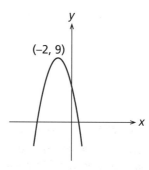

Many quantities in real life are represented by a quadratic expression. In this case, it is often very useful to know their maximum or minimum values.

## Completing the square

*Example*  A ball is thrown upwards. After $t$ seconds, its height $h$ m is given by

$$h = 40t - 5t^2$$

Find the greatest height reached.

Complete the square of the expression.

$$\begin{aligned} 40t - 5t^2 &= -5(t^2 - 8t) \\ &= -5((t^2 - 8t + 16) - 16) \\ &= -5((t-4)^2 - 16) \\ &= 80 - 5(t-4)^2 \end{aligned}$$

The greatest value of this expression is 80, reached when $t = 4$.
The greatest height is 80 m.

## Exercise 3.5

For each of the expressions in questions 1–18, complete the square and find the greatest or least value. Find also the value of $x$ which gives the greatest or least value.

1  $x^2 + 6x + 3$
2  $x^2 + 4x + 7$
3  $x^2 + 10x - 1$
4  $2x^2 + 8x - 7$
5  $3x^2 - 12x + 4$
6  $5x^2 + 5x + 1$
7  $2x^2 + 14x - 3$
8  $x^2 + x - 3$
9  $x^2 - x + 7$
10  $4 + 4x - x^2$
11  $2 - 6x - x^2$
12  $11 + x - x^2$
13  $1 - x - x^2$
14  $3 + 4x - 2x^2$
15  $7 - 10x - 5x^2$
16  $-1 - 4x - x^2$
17  $-5 - 2x - x^2$
18  $-7 + 2x - x^2$

19  A stone is thrown upwards. After $t$ seconds, its height $h$ metres is given by $h = 20t - 5t^2$. Find the greatest height the stone reaches.

20  Suppose a book is priced at £$x$. The marketing manager of the publisher reckons that the number sold will be $C$ copies, where $C = 10\,000 - 1000x$.
   **a** Show that the revenue from the book is £$(10\,000x - 1000x^2)$.
   **b** By completing the square of the expression in part **a**, find the greatest possible revenue from the book.

21  The width of a rectangle is $x$ cm, and its length is $(40 - x)$ cm. Find the greatest possible area of the rectangle.

### Sketching graphs

Once you have completed the square of a quadratic expression, it is easy to sketch its graph. Find where the graph crosses the $y$-axis, then draw the graph through this crossing point so that it has a maximum or minimum at the appropriate place.

*Example* Complete the square of the expression $x^2 - 4x + 1$. Hence sketch the graph of $y = x^2 - 4x + 1$.

By the methods above, $x^2 - 4x + 1 = (x-2)^2 - 3$. Hence there is a minimum at $(2, -3)$. When $x = 0$, $y = 1$. Hence the graph goes through $(0, 1)$.
Now sketch the graph, with a minimum at $(2, -3)$ and passing through $(0, 1)$.

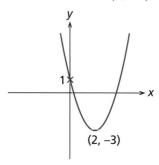

**Note.** The graph of $y = -x^2 + 4x - 1$ will be as shown below, with a maximum at $(2, 3)$ and passing through $(0, -1)$.

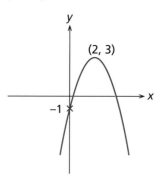

## Exercise 3.6

Sketch the graph of each of the following expressions, by completing the square. If you have a graphical calculator, you can check your answers on it.

1. $y = x^2 - 2x - 1$
2. $y = x^2 - 6x + 2$
3. $y = x^2 + 4x - 1$
4. $y = x^2 + 3x + 5$
5. $y = 1 - 2x - x^2$
6. $y = 3 + 4x - x^2$
7. $y = 2x^2 - 4x + 3$
8. $y = 3x^2 + 12x + 1$
9. $y = 1 - 4x - 2x^2$
10. $y = 7 + 10x - 5x^2$

## Solving equations by completing the square

Here is another reason for completing the square. One advantage of writing a quadratic in completed square form is that there is only one occurrence of $x$, not two. This enables us to solve quadratic equations.

Suppose $x^2 + 8x + 3 = 0$. The expression does not factorise easily, so we cannot solve the equation by factorisation. On page 34 we completed the square of the left-hand side.

## Solving equations by completing the square

$$x^2 + 8x + 3 = (x + 4)^2 - 13$$

Hence $(x + 4)^2 = 13$.
Hence $x + 4 = \pm\sqrt{13}$.
Hence $x = -4 \pm \sqrt{13}$.

> Note that we take both the positive and negative square roots.

*Example* By completing the square, solve the equation $x^2 - 6x - 1 = 0$, giving your answers correct to three decimal places.

Complete the square of the left-hand side.

$$x^2 - 6x - 1 = (x - 3)^2 - 1 - 3^2 = (x - 3)^2 - 10$$

Hence $(x - 3)^2 = 10$.

$$x - 3 = \pm\sqrt{10}$$
$$x = 3 \pm \sqrt{10}$$

> Note that $x - 3$ could be positive or negative. We take both $+\sqrt{10}$ and $-\sqrt{10}$.

To three decimal places, $\sqrt{10} = 3.162$.

$$x = 3 + 3.162 \text{ or } x = 3 - 3.162$$

$x = 6.162$ or $x = -0.162$.

The graph of $y = x^2 - 6x - 1$ is shown. Notice that it crosses the $x$-axis at two points, $-0.2$ and $6.2$. These correspond to the answers above, but are much less accurate.

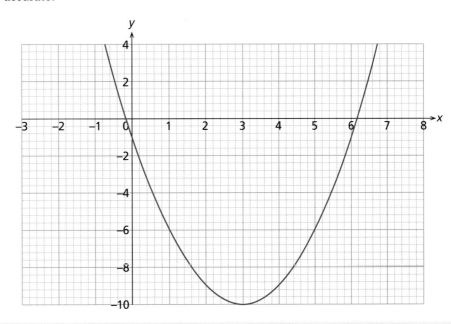

## Exercise 3.7

**1** Solve these equations by completing the square. Give your answers correct to three significant figures.
   **a** $x^2 + 4x - 2 = 0$       **b** $x^2 - 2x - 1 = 0$       **c** $x^2 + 5x + 3 = 0$
   **d** $2x^2 - 4x + 1 = 0$     **e** $3x^2 + 12x - 4 = 0$    **f** $2x^2 + 5x - 1 = 0$

**2** Solve these equations by factorising and by completing the square. Check that your answers are the same.
   **a** $x^2 - 6x + 5 = 0$       **b** $x^2 - 4x - 12 = 0$     **c** $x^2 + 3x - 10 = 0$

## The quadratic formula

The method of completing the square leads to a formula for solving quadratic equations.

- If $ax^2 + bx + c = 0$, then $x = \dfrac{-b \pm \sqrt{b^2 - 4ac}}{2a}$.

**Proof**

Suppose $ax^2 + bx + c = 0$. Divide both sides by $a$.

$$x^2 + \frac{b}{a}x + \frac{c}{a} = 0$$

Complete the square of the left-hand side, by the method of the previous section on page 34. Half the coefficient of $x$ is $\dfrac{b}{2a}$, and when this is squared it becomes $\dfrac{b^2}{4a^2}$.

$$\left(x + \frac{b}{2a}\right)^2 + \frac{c}{a} - \frac{b^2}{4a^2} = 0$$

$$\left(x + \frac{b}{2a}\right)^2 = \frac{b^2}{4a^2} - \frac{c}{a}$$

$$= \frac{b^2 - 4ac}{4a^2}$$

Now take the square root. The result could be either positive or negative.

$$x + \frac{b}{2a} = \pm\sqrt{\frac{b^2 - 4ac}{4a^2}} = \frac{\pm\sqrt{b^2 - 4ac}}{2a}$$

Finally subtract $\dfrac{b}{2a}$ from both sides

$$x = -\frac{b}{2a} + \frac{\pm\sqrt{b^2 - 4ac}}{2a}$$

$$x = \frac{-b \pm \sqrt{b^2 - 4ac}}{2a}$$

This is the formula as required. The next example shows how to use it.

*Example* Use the formula to solve these equations. Give your answers correct to three significant figures.
  **a** $3x^2 + 5x + 1 = 0$  **b** $x^2 = 2x + 5$

**a** Here $a = 3$, $b = 5$ and $c = 1$. Substitute these values into the formula.

$$x = \frac{-b \pm \sqrt{b^2 - 4ac}}{2a}$$

$$= \frac{-5 \pm \sqrt{5^2 - 4 \times 3 \times 1}}{2 \times 3} = \frac{-5 \pm \sqrt{13}}{6}$$

Now use a calculator. $\tfrac{1}{6}(-5 - \sqrt{13}) = -1.43$ and $\tfrac{1}{6}(-5 + \sqrt{13}) = -0.232$.
$x = -1.43$ or $x = -0.232$.

**b** First rearrange the equation so that all the terms are on one side.
$$x^2 = 2x + 5$$
$$x^2 - 2x - 5 = 0$$

The method is similar, but be careful with the negative numbers. $a = 1$, $b = -2$ and $c = -5$.

$$x = \frac{-b \pm \sqrt{b^2 - 4ac}}{2a}$$

$$= \frac{-(-2) \pm \sqrt{(-2)^2 - 4 \times 1 \times (-5)}}{2 \times 1}$$

$$= \frac{2 \pm \sqrt{4 + 20}}{2}$$

$$= \frac{2 \pm \sqrt{24}}{2}$$

> Note that
> $-(-2) = +2$,
> $(-2)^2 = +4$ and that
> $-4 \times (-5) = +20$.

Now use a calculator. $\frac{1}{2}(2 - \sqrt{24}) = -1.45$ and $\frac{1}{2}(2 + \sqrt{24}) = 3.45$.
$x = -1.45$ or $x = 3.45$.

**Note.** You now have four ways of solving a quadratic equation: by factorisation, by graphs, by completing the square and by the formula. The last two methods are superior to the graph method, as they are quicker to use and much more accurate. They are also superior to the factorisation method, as they will always find the solutions when they exist. Indeed, if the quadratic has large coefficients, then it is often better to go straight to the formula rather than try to factorise it. The quadratic $210x^2 + 227x - 1440$ does factorise, but it would take a very long time to find the factors!

# Exercise 3.8

**1** Solve these equations using the formula. Give your answers correct to three significant figures.
  **a** $2x^2 + 5x + 1 = 0$
  **b** $3x^2 + 7x + 1 = 0$
  **c** $x^2 - 4x + 1 = 0$
  **d** $x^2 + 5x - 7 = 0$
  **e** $2x^2 - 5x - 11 = 0$
  **f** $3x^2 - 5x - 3 = 0$

**2** Solve these equations using the formula. Give your answers correct to four significant figures.
  **a** $x^2 = 3x + 5$
  **b** $x^2 + 3 = 6x$
  **c** $x(x + 3) = 8$
  **d** $x(x - 7) = 11$
  **e** $(x + 1)(x - 3) = 14$
  **f** $(x - 2)(x - 5) = 13$
  **g** $x + \frac{1}{x} = 6$
  **h** $x = 5 - \frac{2}{x}$

**3** Here is another way of proving the formula. As above, we want to solve $ax^2 + bx + c = 0$. Multiply all the terms of the equation by $4a$, and compare this with the expansion of $(2ax + b)^2$. Rearrange to make $x$ the subject of the formula.

# Practical uses

Quadratic equations arise in many practical contexts. If the quadratic does not factorise, then you have to solve it by the formula or by completing the square.

*Examples*  A rectangle is 3 m longer than it is wide. If the area is 20 m² find the width.

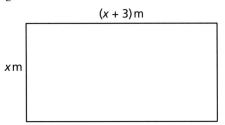

Let the width be $x$ m. Then the length is $(x+3)$ m. As the area is 20 m²,

$$x(x+3) = 20$$
$$x^2 + 3x = 20$$
$$x^2 + 3x - 20 = 0$$

The left-hand side does not factorise. Use the formula, putting $a = 1$, $b = 3$ and $c = -20$.

$$x = -6.217 \text{ or } x = 3.217$$

The negative solution has no meaning here, as a rectangle cannot have a negative width.
The width is 3.22 m.

---

Mrs Jenkins drove a distance of 160 miles from London to Yorkshire. The traffic was heavy, and her average speed was 15 m.p.h. less than expected. The journey took 1 hour longer than expected. What was her speed?

Suppose her average speed was $x$ m.p.h. Without the traffic, the speed would have been 15 m.p.h. faster, i.e. $(x+15)$ m.p.h.

The time of the journey was $\dfrac{160}{x}$ hours. Without the traffic, the time would have been $\dfrac{160}{x+15}$ hours.

The actual time was 1 hour greater than the expected time. Hence

$$\frac{160}{x} = \frac{160}{x+15} + 1$$

Multiply across by $x$ and then by $(x+15)$, and simplify.

$$160 = \frac{160x}{x+15} + x$$

$$160(x+15) = 160x + x(x+15)$$
$$160x + 2400 = 160x + x^2 + 15x$$
$$x^2 + 15x - 2400 = 0$$

Solve by the formula or completing the square.

$$x = 42.1 \text{ or } x = -57.1$$

The negative answer is not relevant in this context.
Her average speed was 42 m.p.h.   (to the nearest whole number).

# Exercise 3.9

1. A ball is thrown up, and $t$ seconds later its height $h$ m is given by $h = 1 + 20t - 5t^2$. When is it 15 m high?
2. A stone is thrown downwards from the top of a tower. After $t$ seconds its has fallen $s$ m, where $s = 10t + 5t^2$. Find when it has fallen 50 m.
3. A man walks $x$ km north then $(x + 8)$ km east. He is now 50 km from his starting point. Find $x$. (Hint: use Pythagoras' theorem.)
4. The width of a rectangle is $x$ cm, and its length is $(x + 15)$ cm. Find the value of $x$ for which the area of the rectangle is 32 cm$^2$.
5. The length of a rectangle is 2 m greater than the width. The area is 27 m$^2$. Let the width be $x$ m. Find an equation in $x$ and solve it.
6. The two sides of a rectangle are $x$ cm and $(x + 3)$ cm. The diagonal of the rectangle is 20 cm. Find $x$.
7. The shorter sides of a right-angled triangle are $x$ cm and $(x - 2)$ cm. The area of the triangle is 4 cm$^2$. Find $x$.
8. The hypotenuse of a right-angled triangle is $(x + 8)$ cm, and the two shorter sides are $(x + 2)$ cm and $(x - 3)$ cm. Find $x$.
9. The area of a hall is 300 m$^2$. If the width is decreased by 1 m and the length increased by 2 m, the area is unchanged. Find the original width.
10. If a cylinder has height $h$ and radius $r$, then its surface area is $2\pi r^2 + 2\pi rh$. Find the radius of a cylinder with height 8 cm and surface area $100\pi$ cm$^2$.
11. A cylinder has height 10 cm and surface area 20 cm$^2$. Find its radius.
12. The shape shown is a square of side $x$ cm with an isosceles triangle at one end. If the total area is 40 cm$^2$, find $x$.

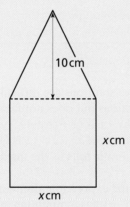

13. The diagram shows a race track, with two straight parts and two semicircles at the ends. The straight parts are $2x$ m apart and they each have length $(2x + 30)$ m. If the area enclosed by the track is 20 000 m$^2$, find $x$.

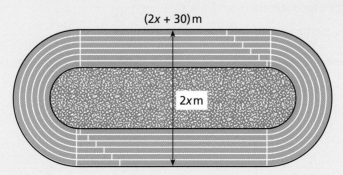

## QUADRATICS

**14** A bus regularly goes on a journey of 60 km. If the speed of the bus is increased by 3 km per hour, the journey will take 0.1 hour less. Find the original speed of the bus.

**15** Keith hikes on a journey of 30 km and back. His average speed for the return journey was 1 km/h greater than for the outward journey. The total time for the hike was 9 hours. Find his speed on the outward journey.

**16** The current in a river is 2 km per hour. A woman can row at $x$ km per hour. She rows 15 km with the current, then 15 km against the current. The total time taken is 8 hours. Find $x$.

**17** The total of the ages in a class is 2002 months. A new student aged 16 years 2 months joins and the average increases by 1 month. Find how many students there were in the class before the new person joined.

**18** A Japanese tourist changed 240 000¥ to £ at a rate of $x$¥ per £. Next day the rate per £ fell by 10¥, and he would have got £40 more if he had waited until then. Find $x$.

**19** An amount of gas has mass 1.2 kg. If its volume increases by 10 m³, then its density decreases by 0.01 kg/m³. Find its original volume.

> **Remember:**
> The density of a substance is its mass divided by its volume.

## SUMMARY

- Some **quadratic equations** which have solutions are difficult to solve by factorisation.
  For example, $x^2 - 2x - 1 = 0$ has solutions but does not factorise easily.
- The **completed square** form of $x^2 + bx + c$ is $(x + \frac{1}{2}b)^2 + c - \frac{1}{4}b^2$.
  For example, the completed square form of $x^2 - 2x - 1$ is $(x - 1)^2 - 2$.
- The completed square form of a quadratic tells us the **maximum** or **minimum** value of the quadratic.
  For example, the minimum value of $x^2 - 2x - 1$ is $-2$, at $x = 1$.
- We can sketch the graph of a quadratic expression after completing the square.
  For example, the graph of $y = x^2 - 2x - 1$ has a minimum at $(1, -2)$.
- Some quadratics can be solved by **completing the square**.
  For example, the solution of $x^2 - 2x - 1 = 1$ is $x = 1 \pm \sqrt{2}$.
- If $ax^2 + bx + c = 0$, then the formula to find $x$ is $x = \dfrac{-b \pm \sqrt{b^2 - 4ac}}{2a}$.
- Quadratic equations can be used to solve practical problems.

## Exercise 3A

1. Factorise $x^2 - 5x - 14$.
2. Solve $x^2 - 11x + 30 = 0$ by factorising.
3. Complete the square of the expression $x^2 + 4x - 7$.
4. Find the maximum or minimum of the expression $x^2 + 4x - 7$.
5. Sketch the graph of $y = x^2 + 4x - 7$.
6. Solve the equation $x^2 + 4x - 7 = 0$.
7. A farmer has 100 m of wire netting to enclose a rectangular sheep pen, one side of which is a long straight wall. Suppose he uses $x$ m for the sides perpendicular to the wall; show that the area enclosed is $x(100 - 2x)$. Complete the square of this expression to find the maximum area he can enclose.

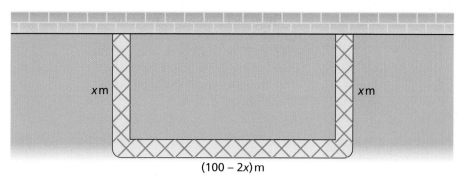

8. Solve the equation $2x^2 - 3x - 10 = 0$ using the quadratic formula.
9. Solve the equation $(x + 3)(x + 4) = 11$ using the quadratic formula.
10. One side of a rectangle is 5 m longer than the other side. The area of the rectangle is 27 m². Find the shorter side of the rectangle.

## Exercise 3B

1. Factorise $x^2 - 8x + 12$.
2. Solve $x^2 - 13x - 30 = 0$ by factorising.
3. Complete the square of the expression $5 - 6x - x^2$.
4. Find the maximum or minimum of the expressions $5 - 6x - x^2$.
5. Sketch the graph of $y = 5 - 6x - x^2$.
6. Solve the equation $5 - 6x - x^2 = 0$.
7. A motorist slows down and then accelerates. After $t$ seconds her speed is $v$ m.p.h., where $v = t^2 - 8t + 40$. Find her least speed during this manoeuvre.
8. Solve the equation $5x^2 + 11x - 13 = 0$ using the quadratic formula.
9. Solve the equation $x + \dfrac{3}{x} = 8$ using the quadratic formula.
10. Andrea walks 10 miles, at $x$ m.p.h. For the return journey, her speed was 1 m.p.h. less. The total time for the journey was 6 hours. Find an equation in $x$ and solve it.

## Exercise 3C

Methods for solving quadratic equations have been known for thousands of years. There are clay tablets, dating from almost 2000 BC, giving quadratic equations and their solutions.

The tablet illustrated, from the British Museum, dates from around 2000 BC. One of its problems and the solution are given below on the left. The modern way of writing it is on the right.

| | |
|---|---|
| I have added the area and the side of my square: 0;45 | $x^2 + x = 0.75$ |
| You write down 1, the coefficient | Coefficient of $x$ is 1 |
| You break half of 1, 0;30 | Half the coefficient of $x$ is 0.5 |
| You multiply 0;30 and 0;30, 0;15 | $(0.5)^2 = 0.25$ |
| You add 0;15 to 0;45, 1 | $0.25 + 0.75 = 1$ |
| This is the square of 1 | $\sqrt{1} = 1$ |
| Subtract 0;30, which you multiplied | $1 - 0.5 = 0.5$ |
| 0;30 is the side of the square. | $x = 0.5$ |

1. The fractions here are sexagesimal, base 60. So 0;45 is $\frac{45}{60} = \frac{3}{4}$. Check the other sexagesimal fractions in the problem.
2. Check that the solution $x = 0.5$ is correct for the original equation.
3. Another problem reads as follows. See if you can complete the modern version on the right.

> We still use sexagesimal fractions in the measurement of time.

| | |
|---|---|
| I have subtracted the side of my square from the area: 870 | $x^2 - x = 870$ |
| You write down 1, the coefficient | |
| You break half of 1, 0;30 | |
| You multiply 0;30 and 0;30, 0;15 | |
| You add 0;15 to 870. Result 870;15 | |
| This is the square of 29;30 | |
| You add 0;30, which you multiplied, to 29;30 | |
| Result 30, the side of the square | |

4. The problems on the tablets are all numerical. But there is a definite rule being followed here. Essentially the rule finds the solution of $x^2 + bx = c$ or of $x^2 - bx = c$. Follow the rule to find formulae for the solution in terms of $b$ and $c$.

## Exercise 3D

The diagram shows a rectangle, *x* units long and 1 unit wide. Suppose a 1 by 1 square is removed from one end. If the remaining rectangle is similar to the original rectangle, then the value of *x* is the **golden ratio**.

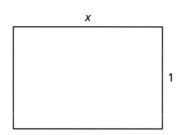

1. Write down the lengths of the sides of the remaining rectangle.
2. Find an equation in *x*.
3. Solve this equation to find the golden ratio.
4. Some artists have used the golden ratio: find a book containing illustrations of Renaissance paintings, and find the ratio of the height to the width of some of them. How close are they to the golden ratio?

> Musicians and architects have also used the golden ratio. Use the internet to find out more about it.

## Exercise 3E

A computer can be used to solve quadratic equations. In particular, you can set up a spreadsheet to solve them.
In A1, B1 and C1 enter the coefficients *a*, *b* and *c* of the equation $ax^2 + bx + c = 0$.
In A2 enter the following formula. (It may vary depending on the spreadsheet package you are using.)

=(-B1+sqrt(B1^2-4*A1*C1))/(2*A1)

This gives the larger solution (assuming *a* is positive). In B2 enter the formula for the smaller solution.

Use this spreadsheet to solve some of the quadratic equations of this chapter.

**You will need:**
- computer with spreadsheet package installed

# 4 Probability

When an event like a football match or a horse race takes place, the different results are called the **outcomes**. One outcome of a soccer match is that the home side loses, and one outcome of a greyhound race is that a particular dog wins. The **probability** of an outcome is a measure of the belief that it will happen.

## Combinations of outcomes

Suppose we have two outcomes that cannot both happen. They are said to be **mutually exclusive**. For example, only one horse can win a given race, so the outcomes of two different horses winning are mutually exclusive. The probability that either horse wins is the sum of the individual probabilities.

P(The Joker wins or Let it Be wins) =
P(The Joker wins) + P(Let it Be wins)

● *In general, if outcomes A and B are mutually exclusive,*
$P(A \text{ or } B) = P(A) + P(B)$.

Suppose we have two events which are not connected in any way. Then the result of the first event will not affect the result of the second event. The result of a football match does not affect the result of a greyhound race. In this case, the outcomes are said to be **independent**. The probability of two outcomes in the events both happening is the product of the individual outcomes.

P(Barchester United loses and Press on Boy wins the race) =
P(Barchester United loses) × P(Press on Boy wins the race)

● *In general, if A and B are independent,* $P(A \text{ and } B) = P(A) \times P(B)$.

## Exercise 4.1

Mainly revision

1. Suppose that The Joker and Let it Be have probabilities $\frac{1}{6}$ and $\frac{1}{8}$ respectively of winning the race. What is the probability that one of the horses wins?
2. In the county cricket championship, Barsetshire has probability $\frac{1}{15}$ of winning and Loamshire has probability $\frac{1}{25}$ of winning. What is the probability that one of these counties will win?
3. A biased dice is such that the probabilities of obtaining 5 and of obtaining 6 in a single throw are $\frac{1}{4}$ and $\frac{1}{5}$ respectively. What is the probability of obtaining 5 or 6?
4. Suppose that the probability that Barchester United wins its football match is 0.4, and the probability that Press on Boy wins its race is 0.08. What is the probability that they both win?
5. In the situation of question 4, find the probability that
   a  neither Barchester United nor Press on Boy wins
   b  Barchester United wins but Press on Boy doesn't
   c  Barchester United doesn't win but Press on Boy does.
6. A four-sided and a six-sided dice are rolled together. What is the probability that
   a  both dice give a 1
   b  neither dice gives a 1.
7. A six-sided dice is rolled and a coin is spun. What is the probability that the dice gives a 6 and the coin gives a Head?

**8** Two boxes hold black and white counters. The first box has 8 white and 7 black counters, and the second box has 5 white and 9 black counters. One counter is drawn from each box. What is the probability they are both white?

**9** In his left pocket Giles has three 5p coins and two 2p coins. In his right pocket he has four 5p coins and two 2p coins. He draws one coin from each pocket. Find the probability that
 **a** they are both 5p coins
 **b** they are both 2p coins.

**10** A menu has 8 choices for starters, of which 3 are vegetarian. There are 12 choices for the second course, of which 5 are vegetarian. A diner picks a starter and a second course at random. What is the probability that he has an entirely vegetarian meal?

**11** In the game of 'Stone, Paper, Scissors' two people each choose at random one of the items stone, paper or scissors. Paper beats stone by wrapping it up, scissors beats paper by cutting it, stone beats scissors by blunting it. The choices are independent of each other. Find the probability that
 **a** both choose stone (so the game is a draw)
 **b** the first person chooses stone and the second scissors (so the first person wins)
 **c** they choose different items.

In general, the outcomes of events are not independent. This chapter deals with combinations of non-independent outcomes.

Kassam has two ways of going to school: walking or by bus. He finds that he is late for school on 10% of the times he walks, and on 20% of the times he goes by bus.

There are two events here: how Kassam decides to go to school, and whether or not he is late. These events are *not* independent. His mode of transport affects whether or not he is late, as he is more likely to be late if he goes by bus. His probability of being late is conditional on his mode of transport.

Suppose two events are *not* independent. Then, if we know $A$ is true, that affects the probability that $B$ is true. In the example above, the probability of Kassam being late depended on how he got to school. Given that he went to school by bus, the probability that he is late is $\frac{1}{5}$. This is the **conditional** probability that he is late.

If two outcomes are not independent, then the probability that they both happen is not the product of the individual probabilities. Instead, we multiply the probability of one event with the *conditional* probability of the other event.

$$P(A \ \& \ B) = P(A) \times P(B, \text{ given that } A \text{ is true})$$

In some cases we can find conditional probabilities.

*Examples* Naomi and Evangeline spin a fair coin to see who will serve first at tennis. The probability that Naomi wins a game in which she serves is $\frac{3}{5}$, and the probability that she wins a game in which Evangeline serves is $\frac{1}{3}$. Find the probability that Naomi serves and wins the first game.

The probability that Naomi wins the spin is $\frac{1}{2}$. She now has probability $\frac{3}{5}$ of winning the game. Multiply these two probabilities together.

$$P(\text{Naomi serves and wins}) = P(\text{Naomi wins spin}) \times P(\text{Naomi wins game, given that she serves}) = \frac{1}{2} \times \frac{3}{5} = \frac{3}{10}$$

The probability that she serves and wins the first game is $\frac{3}{10}$.

# 50 PROBABILITY

A fair coin is spun three times. Find the conditional probability that
**a** there are exactly two Heads, given that the first spin gives Heads
**b** the first spin gave Heads, given that there are exactly two Heads.

**a** The possible outcomes for the second and third spin are

HH   HT   TH   TT

There is already one Head, from the first spin. So, of the four outcomes, two, HT and TH, will give us a total of exactly two Heads.
The probability of exactly two Heads is $\frac{1}{2}$.

**b** The spins of three coins which give rise to exactly two Heads are

HHT   HTH   THH

Of these three outcomes, HHT and HTH have a Head for the first spin.
The probability that the first spin gave Heads is $\frac{2}{3}$.

## Exercise 4.2

1. In the situation of the first example above, find the probability that Evangeline serves and wins.
2. Consider Kassam above. Suppose the probability that he chooses to walk to school is 0.6, and the probability that he takes the bus is 0.4. Find the probability that
   **a** he walks to school and is late
   **b** he takes the bus to school and is on time.
3. Brod's Syndrome is a medical condition that afflicts 5% of the population. There is a test to identify people with the condition, which does not always give the correct results. The test has a positive result for 90% of those with the syndrome and for 20% of those without. A person is picked at random and tested. Find the probability that
   **a** the person has Brod's syndrome and tests positive
   **b** the person doesn't have the syndrome and tests negative.
4. Three-quarters of the pupils in a school were inoculated against 'flu. In the following winter, two thirds of the pupils who hadn't been inoculated caught flu, and a fifth of those who had been inoculated caught 'flu. If a pupil is picked at random, find the probability that
   **a** the pupil had been inoculated and didn't catch 'flu
   **b** the pupil hadn't been inoculated and did catch 'flu.
5. Tony and Gerald play two sets of tennis. The probability that Tony wins the first set is $\frac{2}{5}$. If Tony wins the first set he is encouraged, and the probability that he wins the second is $\frac{3}{5}$. If Gerald wins the first set he gets complacent, and the probability he wins the second is $\frac{1}{2}$.
   **a** Find the probability that Tony wins both sets.
   **b** Find the probability that Gerald wins the first and loses the second.
6. Two fair dice are rolled. Find the conditional probability that
   **a** the total is 8, given that the first dice is a 5
   **b** the total is 8, given that the first dice is a 1
   **c** the first dice is a 3, given that the total is 8
   **d** the first dice is a 5, given that the total is 11
   **e** the first dice is a 6, given that the total is 12.
7. A card is drawn from the standard pack of 52 cards. Find the conditional probability that
   **a** the card is a Diamond, given that it is red
   **b** the card is a King, given that it is a picture card (Jack, Queen or King)
   **c** the card is black, given that it is a Club.

**8** Packets of muesli are either economy size or standard size, and are either 'ordinary' or 'de luxe'. The table below gives the numbers of packets of each type in a shop.

|  | economy size | standard size |
|---|---|---|
| ordinary | 20 | 15 |
| de luxe | 10 | 8 |

One packet is selected at random.
  **a** What is the probability that it is economy size?
  **b** Given that it is economy size, what is the probability that it is de luxe?
  **c** Given that it is de luxe, what is the probability that it is standard size?

**9** The monsters in a computer game are either goblins or trolls, and are either friends or enemies. The table below gives the numbers of each type.

|  | goblin | troll |
|---|---|---|
| friend | 25 | 3 |
| enemy | 5 | 27 |

One of these monsters appears on the screen at random.
  **a** What is the probability that it is a troll?
  **b** Given that it is a goblin, what is the probability that it is friendly?
  **c** Given that it is an enemy, what is the probability that it is a troll?

# Tree diagrams

**Tree diagrams** are a good way to show probabilities. They are particularly useful when the events are not independent. Suppose the probabilities of the outcomes of the second event are dependent on the outcomes of the first event.

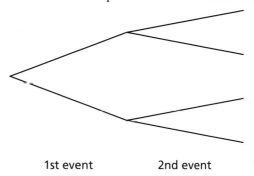

1st event      2nd event

The first fork of the tree diagram leads to branches which give the outcomes of the first event. The next forks of the diagram give the outcomes of the second event. The probabilities of the second outcomes will vary, depending on which branch of the first fork they are attached to.

Consider the situation on page 49, where the probability that Kassam is late depends on how he travelled to school.

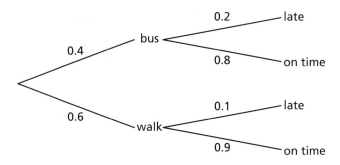

The first fork of the tree diagram corresponds to how he travelled. The second forks correspond to whether or not he was late. The probabilities of the branches from the first fork are the probabilities of the different ways of getting to school, 0.4 and 0.6. The probabilities that he is late depend on how he went to school. They are the *conditional* probabilities that he is late. For the top fork, given that he went by bus, the probabilities are 0.2 and 0.8. For the second fork, given that he walked, they are 0.1 and 0.9.

From the tree diagram we can work out the probability that he is late. This happens on the first and third branches. These have probabilities

P(goes by bus and is late) = 0.4 × 0.2 = 0.08
P(walks and is late) = 0.6 × 0.1 = 0.06

Add these to find the probability that he is late, which is 0.14.

*Examples* Refer to the tennis example on page 49. Find the probability that Naomi wins the first game.

The first fork corresponds to the result of the spin. As it is a fair coin, both branches have probability $\frac{1}{2}$. The next forks correspond to who wins the first game, and the probabilities depend on who serves. For the top fork the probabilities are $\frac{3}{5}$ and $\frac{2}{5}$, and for the bottom fork they are $\frac{1}{3}$ and $\frac{2}{3}$. The completed tree diagram is shown.

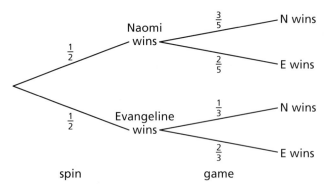

The first and third branches correspond to Naomi winning. Find the probabilities of these branches and add.

$$\tfrac{1}{2} \times \tfrac{3}{5} + \tfrac{1}{2} \times \tfrac{1}{3} = \tfrac{3}{10} + \tfrac{1}{6} = \tfrac{7}{15}$$

The probability that Naomi wins the first game is $\frac{7}{15}$.

*We can add the probabilities of the two branches because they are mutually exclusive.*

**Remember:** *Naomi cannot both win and lose the spin of the coin!*

Dave plays a computer game, in which the monsters are gorgons and gargoyles, in the ratio 3:5. When Dave meets a gorgon, the probability that the gorgon defeats him is $\frac{1}{10}$.

**a** Find the probability that the first monster he meets is a gorgon and that he is defeated by it.

**b** The probability that Dave will be defeated by the first monster he meets is $\frac{1}{12}$. Find the probability that when Dave meets a gargoyle it defeats him.

Set up a tree diagram as shown. The probabilities that the first monster is a gorgon or a gargoyle are $\frac{3}{8}$ and $\frac{5}{8}$ respectively. A gorgon has probability $\frac{1}{10}$ of defeating Dave. Let the probability that a gargoyle defeats Dave be $p$.

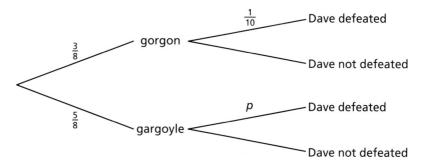

**a** Multiply the probabilities along the top branch.
The probability that Dave meets a gorgon first and is defeated is $\frac{3}{80}$.

**b** The probability that Dave is defeated by the first monster he meets is found from the first and third branches. The sum of these probabilities is $\frac{1}{12}$.

$$\tfrac{3}{8} \times \tfrac{1}{10} + \tfrac{5}{8} \times p = \tfrac{1}{12}$$
$$\tfrac{5}{8} \times p = \tfrac{1}{12} - \tfrac{3}{8} \times \tfrac{1}{10} = \tfrac{11}{240}$$
$$p = \tfrac{8}{5} \times \tfrac{11}{240}$$

When Dave meets a gargoyle, the probability that it defeats him is $\frac{11}{150}$.

## Exercise 4.3

**1** In the tennis example on page 49, find the probability that the server wins the first game.

**2** Refer to the 'flu example in question 4 of exercise 4.2.
  **a** Copy and complete the tree diagram showing the possible outcomes.
  **b** What is the probability that the pupil did catch 'flu?

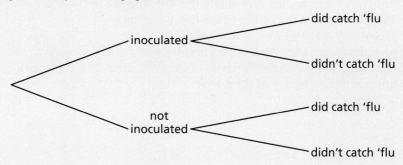

3 On a fine day, the probability that Jennifer will walk to school is 0.4. If the day isn't fine, the probability falls to 0.05. Two out of three days are fine.
   a Copy and complete the tree diagram showing the possible outcomes.
   b What is the probability that it is fine and Jennifer walks to school?
   c What is the probability that she walks to school?

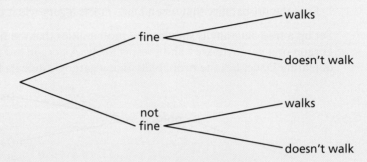

4 The electorate of a constituency is divided between people under thirty and people over thirty in the ratio 3:5. The Purple Party is supported by 30% of the people under thirty and by 40% of the people over thirty. A voter is picked at random.
   a Draw a tree diagram to show the possible outcomes.
   b What is the probability that the voter is under thirty and supports the Purple Party?
   c What is the probability that the voter supports the Purple Party?

5 A factory has two machines making a certain component. Machine A produces twice as many components as machine B. The probabilities that a component made by machine A or machine B is faulty are $\frac{1}{20}$ and $\frac{1}{10}$ respectively. One component is selected for inspection.
   a Draw a tree diagram to show the possible outcomes.
   b What is the probability that the component was made by machine A and is faulty?
   c What is the probability that the component is not faulty?

6 The sixth form of a school contains 80 girls and 70 boys. A level maths is studied by 20% of the girls and 30% of the boys. A student is selected at random.
   a Draw a tree diagram showing the possible outcomes.
   b What is the probability that the student is a girl studying maths?
   c What is the probability that the student does not study maths?

7 Leroy can choose to study either French or German for GCSE (he cannot study both). He is twice as likely to pick French as German. He reckons that his probability of getting an A grade is $\frac{1}{3}$ for French, and $\frac{1}{5}$ for German.
   a Draw a tree diagram to show the possible outcomes.
   b What is the probability that Leroy chooses French and gets an A?
   c What is the probability that he gets an A?

8 Mel and Norma play squash, until one of them has won two games (i.e. the best of three). The probability that Mel wins the first game is 0.6. If she wins a game, the probability that she wins the next is 0.7. If Norma wins a game, the probability that she wins the next is 0.6.
   a Draw a tree diagram showing the results of the first two games.
   b What is the probability that Mel wins in two games?
   c What is the probability that the match is over in two games?
   d Extend your tree diagram to include the result of the third game (if it is played).
   e What is the probability that Mel wins the match?

9 Three people, Peter, Mike and Yan, want to play squash (at which only two people can play). They 'spin out' to see who will play. Each spins a coin. If the coins are all the same, they spin again. If one person's coin is different from the others, then that person sits out.
  a Complete the tree diagram shown.
  b What is the probability that they will have to spin more than once?
  c What is the probability that Peter will play after one round of spinning coins?

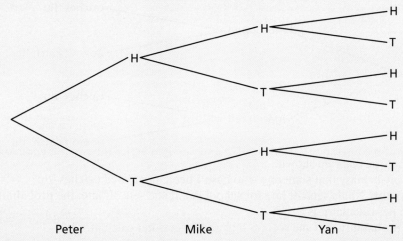

10 Sean and Marie play the game of 'Stone, Paper, Scissors' of question 11 of exercise 4.1.
  a Complete the tree diagram shown.
  b What is the probability that the game is a draw?
  c What is the probability that Sean wins the game?

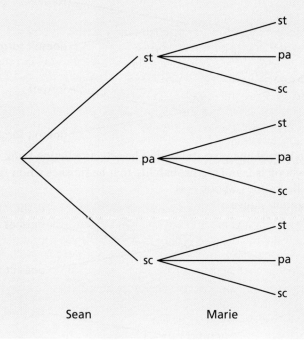

11 Two thirds of the pupils in a school have been inoculated against 'flu. If someone has been inoculated, the probability that he or she will catch 'flu is $\frac{1}{15}$.
  a If a pupil is picked at random, what is the probability that he or she has been inoculated and catches 'flu?
  Overall, a fifth of the pupils catch 'flu. Let the probability that someone who hasn't been inoculated catches 'flu be $q$.

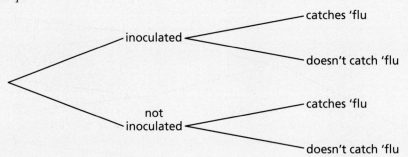

  b Complete the tree diagram shown.
  c What is the probability that someone who hasn't been inoculated catches 'flu?

12 The probability that Nuala sets off late for school is $\frac{1}{6}$. If she sets off late, the probability that she forgets to take her calculator is $\frac{1}{4}$.
  a What is the probability that she sets off late and forgets her calculator?
  Overall, the probability that she forgets her calculator is $\frac{1}{10}$. Let the probability that she sets off on time and forgets her calculator be $p$.
  b Complete the tree diagram shown.
  c If she sets off on time, what is the probability that she remembers her calculator?

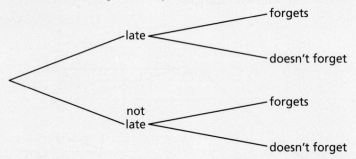

13 When Bryan goes to the library, the probability that he selects a fiction book is $p$. The probability that he finishes a fiction book is $\frac{3}{4}$, and the probability that he finishes a non-fiction book is $\frac{2}{3}$. The overall probability that he finishes a book is $\frac{43}{60}$.
  a Complete the tree diagram shown.
  b Find $p$.

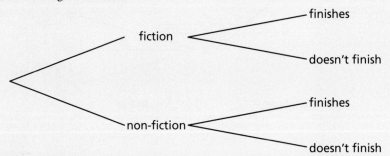

14 A driving test centre tests car drivers and motorcycle riders. The pass rates for car drivers and motorcycle riders are $\frac{7}{9}$ and $\frac{5}{8}$ respectively. The overall pass rate is $\frac{17}{24}$. Find the proportion of car drivers at the centre.

# Drawing with and without replacement

Consider these two situations. In both cases I have a bag containing 4 red counters and 6 blue counters, and I draw (i.e. take) two counters from the bag.

> I draw the first counter, put it back in the bag and then draw the second counter.
> I draw the first counter, put it aside and then draw the second counter.

The first situation is called **drawing with replacement**. The counter is replaced in the bag before the second draw is made.

The second situation is called **drawing without replacement**. The counter is not replaced in the bag before the second draw is made.

The probability of drawing two red counters depends on whether the drawing is with or without replacement.

Suppose the drawing is with replacement. Then the probability that the second counter is red is always $\frac{4}{10}$, regardless of the result of the first drawing. The tree diagram shows the possible outcomes from the two drawings.

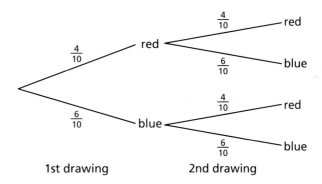

The probability of drawing two red counters is the probability of the top branch.

$$P(\text{two reds}) = \tfrac{4}{10} \times \tfrac{4}{10} = \tfrac{16}{100} = \tfrac{4}{25}$$

Suppose the drawing is without replacement. Then the probability that the second counter is red depends on the result of the first drawing. If the first counter is red, then there are now 3 red counters out of 9. So the probability that the second counter is red is $\frac{3}{9}$, i.e. $\frac{1}{3}$. If the first counter is blue, there are now 4 red counters out of 9, so the probability that the second counter is red is $\frac{4}{9}$.

A tree diagram for the outcomes of the two drawings is shown.

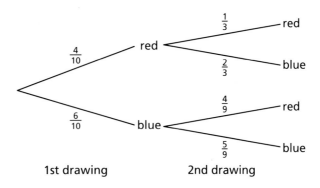

The probability that both counters are red is the product of the probabilities along the top branch.

$$P(\text{both red}) = \tfrac{4}{10} \times \tfrac{1}{3} = \tfrac{4}{30} = \tfrac{2}{15}$$

So, if the drawing is with replacement, the outcomes of the two drawings are independent of each other. If the drawing is without replacement, the outcomes are not independent.

*Example* Two cards are drawn without replacement from a standard pack of cards. What is the probability that they are both Hearts?

The probability that the first card is a Heart is $\tfrac{13}{52}$, i.e. $\tfrac{1}{4}$. If the first card drawn is a Heart, there are now 12 Hearts in a pack of 51, so the probability that the second card is a Heart is $\tfrac{12}{51}$, which is $\tfrac{4}{17}$.

To find the probability that both cards are Hearts multiply these probabilities together.

$$\tfrac{1}{4} \times \tfrac{4}{17} = \tfrac{1}{17}$$

The probability that both cards are Hearts is $\tfrac{1}{17}$.

> If the drawing was with replacement, then the probability that both cards are Hearts would be $\tfrac{1}{4} \times \tfrac{1}{4}$, i.e. $\tfrac{1}{16}$. This probability is slightly greater than $\tfrac{1}{17}$.

# Exercise 4.4

1 A box contains 5 white and 7 black counters. Two are drawn. Find the probability that both are black when
   a the drawing is with replacement
   b the drawing is without replacement.
2 The letters A, B, C, D, E, F, G are written on pieces of paper which are put into a hat. Two are drawn. Find the probability that both are vowels when
   a the drawing is with replacement
   b the drawing is without replacement.
3 A sample of voters contains 10 men and 12 women. A voter is selected at random twice. Find the probability that both are women if
   a the same voter can be selected twice
   b the voters must be different.
4 A pile of books contains 7 novels and 6 biographies. Two are selected at random (without replacement).

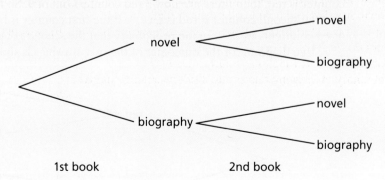

   a Copy and complete the tree diagram shown.
   b What is the probability that both books are novels?
   c What is the probability that exactly one book is a novel?
   d What is the probability that at least one book is a novel?

5 Two cards are dealt from the top of a standard pack.

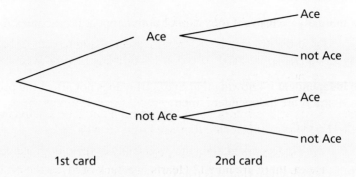

  a  Copy and complete the tree diagram shown.
  b  What is the probability that both cards are Aces?
  c  What is the probability that exactly one is an Ace?
  d  What is the probability that at least one is an Ace?
6 The numbers 1 to 9 are written on pieces of paper which are put in a hat. Two are drawn without replacement.

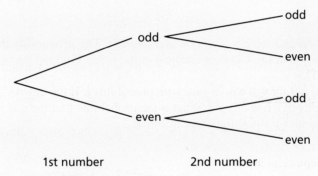

  a  Copy and complete the tree diagram shown.
  b  What is the probability that the product of the numbers is odd?
  c  What is the probability that the sum of the numbers is odd?
7 A bag contains 4 red and 3 green marbles. Two are drawn without replacement.
  a  Draw a tree diagram showing the possible results.
  b  What is the probability that both marbles are green?
  c  What is the probability that one is red and one is green?
8 A box contains 5 round counters and 7 square counters. Two are drawn without replacement.
  a  Draw a tree diagram to show the possible results.
  b  What is the probability of drawing two square counters?
  c  What is the probability of drawing one round and one square counter?

## SUMMARY

- Two **outcomes** are **mutually exclusive** if they cannot both happen. If outcomes $A$ and $B$ are mutually exclusive, then
  $$P(A \text{ or } B) = P(A) + P(B)$$
- Two outcomes are **independent** if knowing that one is true does not affect the probability of the other. If outcomes $A$ and $B$ are independent, then
  $$P(A \& B) = P(A) \times P(B)$$
- If outcomes $A$ and $B$ are *not* independent, then the probability of them both happening can be found by
  $$P(A \& B) = P(A) \times P(B, \text{ given that } A \text{ is true})$$
  i.e. the probability of $A$ multiplied by the **conditional** probability of $B$.
- A **tree diagram** can be used to find probabilities of combinations of non-independent outcomes.
- When **drawing with replacement**, the item is put back after selection. When **drawing without replacement**, the item is not put back. If drawing is with replacement, the outcomes on the first and second drawings are independent.

## Exercise 4A

1. Janice goes to the library, and selects one book and one CD. The probability that the book is a novel is 0.7, and the probability that the CD is of classical music is 0.3. Find the probability that she selects a novel and a classical CD.
2. On a fine day, the horse al Jabr will win its race with probability $\frac{1}{5}$. If the day isn't fine, the probability drops to $\frac{1}{8}$. The probability of a fine day is $\frac{3}{4}$. What is the probability that the day is fine and al Jabr wins?
3. Complete the tree diagram for the situation of question 2.

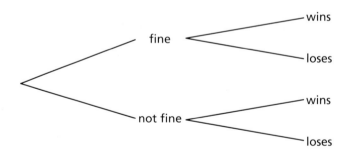

4. Use your tree diagram of question 3 to find the probability that al Jabr wins.
5. Two tetrahedral (four-sided) dice have the numbers 1 to 4 on their faces. They are both rolled and the numbers they land on are noted.
   a. If the first lands on number 2, what is the probability that the total is 5?
   b. If the total is 4, what is the probability that the first dice landed on 2?
6. A panel of voters contain 20 men and 25 women. Of the men, 12 are over 40, and of the women, 15 are over 40. One voter is picked at random. Given that this voter is over 40, what is the probability that he or she is a woman?
7. The squad for a cricket team contains 13 batsmen, of whom 3 are left-handed. Two are picked at random to be the opening pair for a match.
   a. Draw a tree diagram to show the possible outcomes.
   b. What is the probability that both batsmen are left-handed?
   c. What is the probability that one is left-handed and one right-handed?

## Exercise 4B

**1** A roulette wheel has holes numbered 1 to 36. In two spins of the wheel, what is the probability that an even number comes up for the first spin, and a number divisible by 3 for the second spin?

**2** A number is drawn at random from 1 to 19 inclusive.
  **a** Given that the number is even, what is the probability that it is prime?
  **b** Given that the number is prime, what is the probability that it is odd?

**3** A vending machine sells cups of coffee which are either black or white, and which are either with sugar or without. The table below gives the numbers sold in a morning.

| black with sugar | black without sugar | white with sugar | white without sugar |
|---|---|---|---|
| 24 | 38 | 43 | 52 |

For a cup picked at random,
  **a** given that it is black, what is the probability that it is with sugar
  **b** given that it is without sugar, what is the probability that it is white?

**4** Samantha will enter either the 100 m or the 200 m race. She cannot enter both. She is three times more likely to enter the 100 m as the 200 m, because she reckons her probability of winning the 100 m is $\frac{1}{2}$, while her probability of winning the 200 m is only $\frac{1}{6}$. What is the probability that she enters the 100 m and loses?

**5** Draw a tree diagram to show the possible outcomes of the situation of question 4.

**6** Use your tree diagram for question 5 to find the probability that Samantha wins the race she chooses.

**7** In his drawer, Charles has 7 black and 8 white socks. Dressing hurriedly in the dark, he pulls out two at random.
  **a** Draw a tree diagram to show the possible selections.
  **b** What is the probability that he has a pair of the same colour?

## Exercise 4C

**You will need:**
- calculator OR
- computer with spreadsheet package installed

How many people are there in your class? Do you all have different birthdays, or are there at least two people with the same birthday? With a class of 25 or so it is more than likely that at least two people share a birthday.

To find the probability that at least two people share a birthday, find the probability that everyone has a *different* birthday and subtract from 1.

It doesn't matter what the first person's birthday is. The probability that the second person has a different birthday from the first is $\frac{364}{365}$. The probability that the third person also has a different birthday from the first two is $\frac{364}{365} \times \frac{363}{365}$.

Continue, until you have reached the number of people in your class. This requires careful use of a calculator. Alternatively, you could set up a spreadsheet to evaluate the probability. (Put the numbers 365, 364, 363 etc. in the A column. Put 1 in cell B1, then in cell B2 enter =B1*A2/365. Copy this formula down the B column.)

## Exercise 4D

In many of the examples of this chapter you were able to work out the conditional probability. Often, though, it has to be found by experiment. When trying to establish a connection between an item of diet and a medical condition, researchers try to show that the probability of contracting the condition is greater for those who eat the item than for those who don't.

Find a conditional probability by experiment. Here are some suggestions.

- Given that a person is left-handed, what is the probability that he or she wears glasses?
- Given that a football team is playing at home, what is the probability that it wins?

## Exercise 4E

Suppose a fair coin is spun many times. The 'law of large numbers', often called the 'law of averages', states that the proportion of Heads will probably settle down to a value close to $\frac{1}{2}$. You can verify this law on a computer. A spreadsheet is ideal for this.

**You will need:**
- computer with spreadsheet package installed

> The 'law of averages', or Bernoulli's theorem, was described by the Swiss mathematician Jacob Bernoulli (1655–1705) in his *Ars Conjectandi*, published posthumously in 1713.

Number each go using the A column. Enter 1 in A1, 2 in A2 and so on up to A100, or further if you want. (To do this, you can use the 'fill' command, or you can enter =A1+1 in A2 and copy down to A100.)

We want a function which returns 0 (Tail) and 1 (Heads) with equal probability. This is =INT(RAND()*2), this may vary depending on the spreadsheet package you are using. Enter this in B1 and copy down to B100.

The C column will contain the total of Heads. In C1 enter =B1. In C2 enter =B2+C1, and copy down to C100.

The D column will contain the proportion of Heads. In D1 enter =C1/A1 and copy down to D100.

Describe what happens to the numbers in the D column. You can repeat the experiment by 'recalculating'. Do this several times. What is the range of values of the proportion of Heads in 100 spins?

Plot a graph of the D column against the A column – it might look like this.

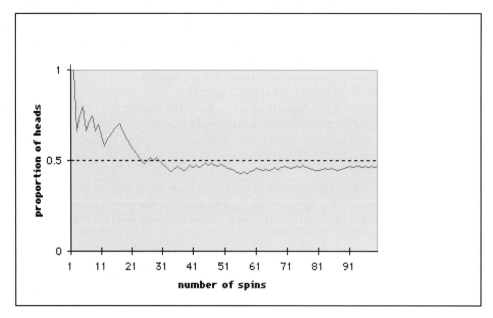

You can also use the spreadsheet to record the result of an experiment to find a probability which cannot be found by theory. It might be that you are finding by experiment the probability that a drawing pin will land point upwards. Repeat the experiment many times, and enter 1 in the B column if the pin lands point upwards, and 0 otherwise. The D column records the relative frequency of landing point upwards. Does it settle down to a fixed value?

# 5 Circles

Circles are simple objects, or so it seems. But there are a great number of unexpected results about them. This chapter contains a sequence of these results, sometimes called **circle theorems**.

## Basic circle theorems

Here are some of the circle theorems you will have met before.

### Two tangent theorems

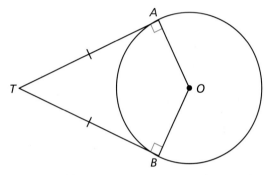

Suppose tangents are drawn from a point $T$ to a circle with centre $O$. Let the tangents touch the circle at $A$ and $B$. Then

$TA = TB$         (tangents from a point are equal)
$\angle TAO = 90°$     (the tangent is perpendicular to the radius)

**Note.** In the quadrilateral $TAOB$, $OA = OB$ (both radii of the circle) and $TA = TB$. So $TAOB$ has two pairs of adjacent sides equal, and hence it is a kite. It is called the **tangent kite**.

### Bisector of a chord theorem

Suppose $AB$ is a chord of a circle. Then the perpendicular bisector of the chord goes through the centre of the circle.

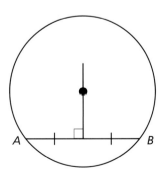

## Angle in a semicircle

Suppose $AB$ is a diameter of a circle, and $C$ is a point on the circumference. Then
$$\angle ACB = 90°$$

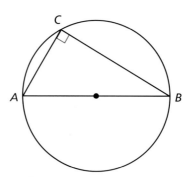

We can use these results to find lengths and angles within circles.

*Examples*   The tangents from $T$ to a circle touch it at $A$ and $B$. If $\angle ATB = 56°$, find $\angle ABT$.

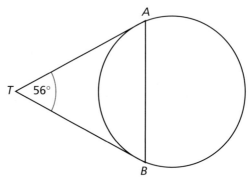

From the first result above, $TA = TB$. Hence $\triangle ATB$ is isosceles, and so $\angle ABT = \angle TAB$.
$$\angle ABT = \tfrac{1}{2}(180° - 56°)$$
$\angle ABT = 62°$.

---

$TA$ is a tangent from $T$ to a circle with centre $O$. If $TO = 20\,\text{cm}$ and $TA = 16\,\text{cm}$ find the radius of the circle.

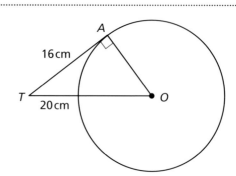

From the third result above, $\angle TAO = 90°$. Using Pythagoras' theorem
$$TO^2 = TA^2 + AO^2$$
$$AO^2 = 20^2 - 16^2 = 144$$

The radius of the circle is $12\,\text{cm}$.

A horizontal drain pipe has radius 6 cm. There is water in it, to a depth of 2 cm. What is the width of the surface of the water?

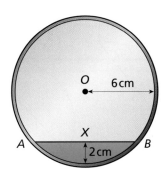

The surface of the water is 4 cm below the centre of the pipe. In the diagram, $X$ is the midpoint of $AB$, at the centre of the water. By the second result above, $XO$ is perpendicular to the surface of the water. Using Pythagoras' theorem

$$XA = \sqrt{6^2 - 4^2} = 4.47$$

The width of the surface of the water is 8.9 cm.

---

$AB$ is a diameter of a circle, and $C$ is a point on the circumference. If $AB = 10$ cm and $AC = 7$ cm, find $\angle ABC$.

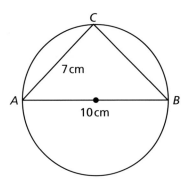

By the third result above, $\angle ACB = 90°$. Using trigonometry in $\triangle ABC$

$$\sin \angle ABC = \tfrac{7}{10}$$

$\angle ABC = 44.4°$.

# Exercise 5.1

Mainly revision

**1** Find the unknown angles in these diagrams. In each case $O$ is the centre of the circle.

a

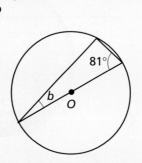

b

c

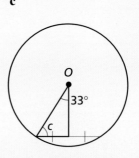

**2** $MN$ is a diameter of a circle and $L$ is a point on the circle. If $\angle MNL = 35°$, find $\angle NML$.

**3** $AB$ and $CD$ are diameters of a circle. Show that $ACBD$ is a rectangle.

**4** $AB$ is a diameter of a circle, radius 10 cm, and $C$ is a point on the circumference. If $CB = 12$ cm, find $CA$.

**5** $AB$ is a diameter of a circle, radius 16 cm, and $C$ is a point on the circumference. If $CB = 7$ cm, find $\angle CAB$.

**6** $AB$ is a diameter of a circle, radius 40 cm, and $C$ is a point on the circumference. If $\angle CBA = 62°$, then find $CA$.

**7** $PQ$ is a diameter of a circle and $R$ is a point on the circumference. $\angle QPR = 37°$. Find $\angle PQR$.

**8** $TA$ is a tangent to a circle of centre $C$. If $TC = 13$ cm and $TA = 10$ cm, find the radius of the circle.

**9** $XA$ is a tangent to a circle of centre $C$ and radius 5 cm. If $XA = 12$ cm, find $XC$.

**10** A chord of length 9 cm is in a circle of radius 5 cm. Find the distance of the chord from the centre of the circle.

**11** In the diagram a quadrilateral $ABCD$ is drawn round a circle. Show that $AB + CD = AD + BC$.

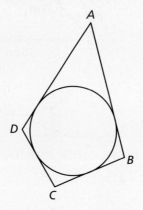

**12** A circular mirror has radius 20 cm. It hangs from a peg 25 cm above the centre of the mirror, by a string which passes round the mirror as shown. Find the total length of the string.

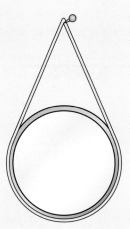

Next we are going to look at a sequence of theorems about angles in a circle. First we have a practical exercise.

# Exercise 5.2

The diagram shows a circle with centre $O$. Points $A$, $C$, $D$, $B$ and $E$ are on the circumference, and $AEF$ is a straight line. Use your protractor to measure the angles $\angle AOB$, $\angle ACB$, $\angle ADB$, $\angle AEB$ and $\angle FEB$.

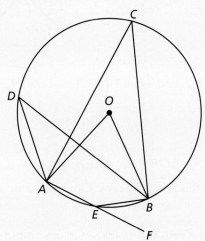

You should find that

$\angle AOB = 2 \times \angle ACB$
$\angle ACB = \angle ADB$
$\angle ACB + \angle AEB = 180°$
$\angle FEB = \angle ACB$

These four results are cases of four theorems.

## Angle at the centre

### Theorem 1

The diagram shows a circle with centre $O$, and a chord $AB$. $C$ is a point on the circumference, on the major arc. The theorem states:

- *The angle subtended by the chord at the centre is twice that at the circumference. In symbols, $\angle AOB = 2 \times \angle ACB$.*

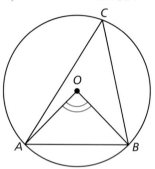

**Proof**

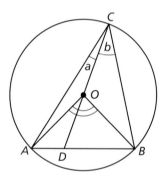

Join $C$ to $O$, and extend it to $D$ as shown. Let $\angle ACO = a$ and $\angle BCO = b$. Then

$\angle CAO = a$ and $\angle CBO = b$     ($\triangle ACO$ and $\triangle BCO$ are isosceles)
$\angle AOD = 2a$ and $\angle BOD = 2b$     (external angles of triangles)

$\angle ACB = a + b$ and $\angle AOB = 2a + 2b = 2(a + b)$.
This gives $\angle AOB = 2 \times \angle ACB$, as required.

**Note.** In this result we took $C$ on the major arc. The theorem still holds if $C$ is on the minor arc as shown here. In this case the *reflex* angle $AOB$ is twice $\angle ACB$.

The theorem also holds if the lines of the diagram cross each other as in the second diagram.

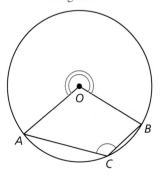 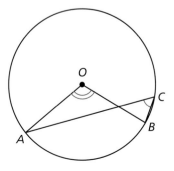

*Example* In the diagram, $O$ is the centre of the circle. Find the angle $\angle OAB$.

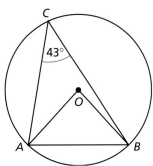

From Theorem 1,
$$\angle AOB = 2 \times \angle ACB$$
$$= 2 \times 43° = 86°$$

$OAB$ is an isosceles triangle, hence $\angle OAB = \angle OBA$.

$$\angle OAB = \tfrac{1}{2} \times (180° - 86°) = \tfrac{1}{2} \times 94°$$

$$\angle OAB = 47°.$$

# Exercise 5.3

You will need:
- compasses
- ruler
- protractor

1 Verify Theorem 1 by an accurate drawing. Draw a circle with centre $O$, draw a chord $AB$ and pick $C$ on the circumference on the major arc. By measurement verify that $\angle AOB = 2 \times \angle ACB$.

2 Find the unknown angles in these diagrams. In each case $O$ is the centre of the circle.

**a**

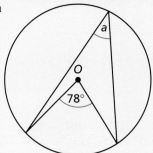

**b**

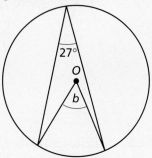

**c**

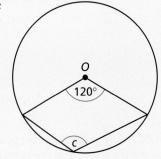

3 If $x = 61°$, find $y$. (The diagram is not drawn accurately.)

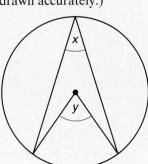

**4** If $b = 146°$, find $a$. (The diagram is not to scale.)

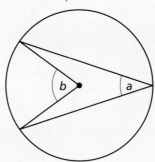

**5** In the diagram, $O$ is the centre of the circle. $\angle OAC = 38°$ and $\angle OBC = 24°$. Find $\angle AOB$.

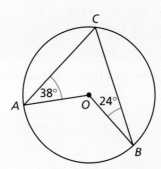

**6** $AB$ is a chord of a circle with centre $O$. $C$ is a point on the major arc. If $\angle OAB = 57°$, find $\angle ACB$.

**7** In the diagram, $O$ is the centre of the circle and $OA = OB = AB$. Find $\angle BCA$.

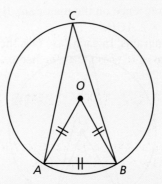

**8** $ABCDE$ is a regular pentagon inscribed in a circle. Calculate $\angle BEC$.

**9** In the diagram, $X$ is the centre of the circle and $\angle ATB = 46°$. Find $\angle ACB$.

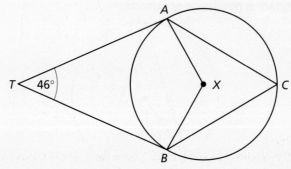

**10** $ABC$ is inscribed in a circle of centre $O$. If $\angle AOB = 126°$ and $\angle BOC = 136°$, find the angles of $ABC$.

11 In the proof of Theorem 1 we took *C* on the major arc. The proof still holds if *C* is on the minor arc. In this diagram prove that the *reflex* angle *AOB* is twice ∠*ACB*.

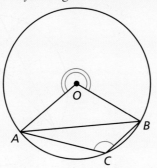

You have already met a special case of Theorem 1. If *AB* is a diameter of the circle, then *AOB* is a straight line. Hence ∠*AOB* = 180°. If *C* is a point on the circumference, then ∠*ACB* = ½ × 180°, which is 90°.

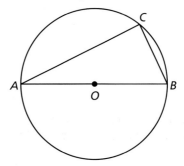

● *The angle in a semicircle is a right angle.*

This was one of the results at the beginning of this chapter.

# Angles on the same arc
### Theorem 2

● *Angles on the same arc are equal. If AB is a chord, and C and D are on the same arc, then ∠ACB = ∠ADB.*

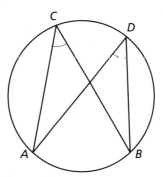

**Proof**
This follows directly from Theorem 1. If *O* is the centre of the circle, then ∠*AOB* = 2 × ∠*ACB* = 2 × ∠*ADB*. It follows that ∠*ACB* = ∠*ADB*.

*Example* In the diagram, $\angle CAB = 63°$ and $\angle CBA = 59°$. Find $\angle ADB$.

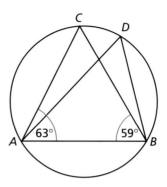

By subtraction from 180°, $\angle ACB = 58°$.

$\angle ADB = \angle ACB$ (angles on the same arc)

Hence $\angle ADB = 58°$.

## Exercise 5.4

You will need:
- compasses
- ruler
- protractor

1 Verify Theorem 2 by an accurate drawing. Draw a circle and draw a chord $AB$. Pick points $C_1$, $C_2$, $C_3$, on the circumference on the same arc, and verify that $\angle AC_1B = \angle AC_2B = \angle AC_3B$.

2 Find the unknown angles in these diagrams.

a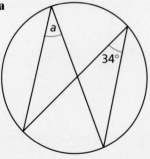

b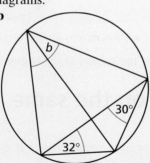

c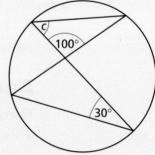

3 Make a copy of these diagrams, and find all the unmarked angles.

a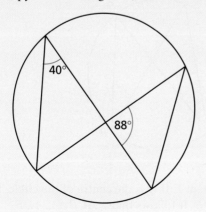

b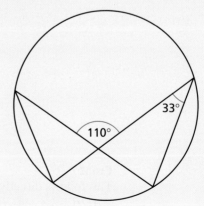

**4** This diagram is not drawn accurately.

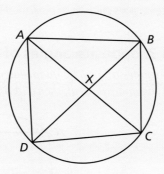

  **a** If ∠ABD = 35° and ∠BDC = 47°, find ∠CXD.
  **b** If ∠AXD = 72° and ∠DBC = 48°, find ∠BDA.
**5** In the diagram, ∠EAB = 32° and ∠AED = 41°. Find ∠BCE.

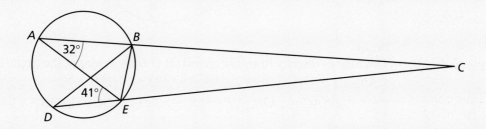

**6** In the diagram, AB is parallel to CD. If ∠CBA = 43°, find ∠AXC.

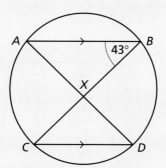

**7** In the diagram, show that the triangles ABX and CDX are similar.

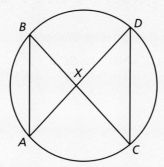

# Cyclic quadrilaterals

A quadrilateral is **cyclic** if it is inscribed in a circle, i.e. if all the vertices are on the circle.

### Theorem 3: opposite angles of a cyclic quadrilateral

- *Opposite angles of a cyclic quadrilateral add up to 180°. If ABCD are the points of the quadrilateral, then $\angle ABC + \angle ADC = 180°$.*

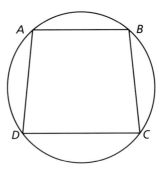

**Proof**

This follows directly from Theorem 1. If $O$ is the centre of the circle, then $\angle ABC = \frac{1}{2} \times \angle AOC$. Also $\angle ADC = \frac{1}{2} \times$ (reflex $\angle AOC$). So

$$\angle ABC + \angle ADC = \tfrac{1}{2}(\angle AOC + \text{reflex } \angle AOC) = \tfrac{1}{2} \times 360° = 180°$$

as required.

### Theorem 4: external angle of a cyclic quadrilateral

- *The exterior angle of a cyclic quadrilateral is equal to the opposite angle. In the diagram, $\angle DAB = \angle BCE$.*

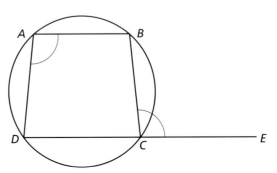

**Proof**

This follows directly from Theorem 3.

$$\begin{aligned}\angle BCE &= 180° - \angle BCD && \text{(angles in a straight line)} \\ &= 180° - (180° - \angle DAB) && \text{(angles in a cyclic quadrilateral)} \\ &= \angle DAB\end{aligned}$$

# Cyclic quadrilaterals

*Example* In the diagram, $O$ is the centre of the circle. If $\angle AOC = 130°$, find $\angle ADC$.

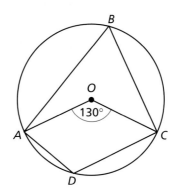

$\angle ABC = \frac{1}{2}\angle AOC$ (angle at centre)
$\quad\quad\quad = 65°$
$\angle ADC = 180° - \angle ABC$ (opposite angles of a cyclic quadrilateral)
$\quad\quad\quad = 180° - 65°$

Hence $\angle ADC = 115°$.

## Exercise 5.5

**1** Find the unknown angles in these diagrams. $O$ is the centre of the circle.

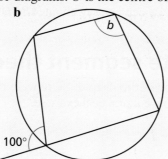

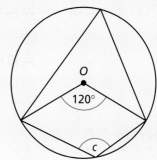

**2** Copy these diagrams, and fill in the unknown angles.

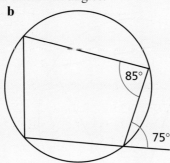

**76** CIRCLES

3  In the diagram, *ACF* and *BDE* are straight lines. $\angle BAC = 110°$. Find $\angle CFE$. What can you say about the lines *AB* and *FE*?

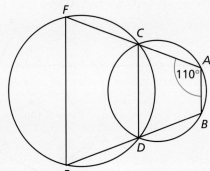

4  In the diagram, $\angle ABC = 80°$ and $\angle BCD = 110°$. Find $\angle CED$.

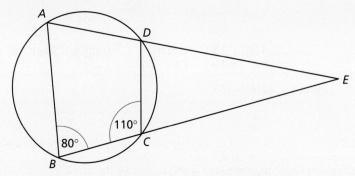

5  *ABCD* is a cyclic parallelogram. What else can you say about *ABCD*?

## Alternate segment theorem

At the beginning of this chapter there were two results about tangents. Here is a third. First we have a practical exercise.

### Exercise 5.6

The diagram shows a circle, and *AT* is the tangent at *A*. Measure $\angle TAB$ and $\angle ACB$. You should find that they are equal.

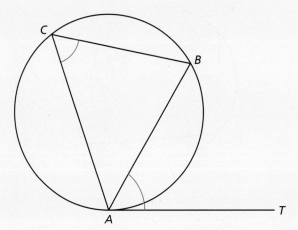

Exercise 5.6 is a case of the next theorem.

### Theorem 5: alternate segment

Suppose a circle has a tangent $TA$ touching it at $A$, $AB$ is a chord, and $C$ is a point in the other (alternate) segment.

- *The angle between the tangent and the chord is equal to the angle in the alternate segment.*

$$\angle TAB = \angle ACB$$

**Proof**

Let $O$ be the centre of the circle. Join $O$ to $A$ and $B$ as shown. Let $\angle TAB = x$.

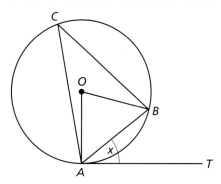

$\angle OAT = 90°$                                      (tangent and radius are perpendicular)
$\angle OAB = 90° - x$
$\angle OBA = \angle OAB = 90° - x$
$\angle AOB = 180° - (90° - x) - (90° - x) = 2x$
$\angle ACB = \tfrac{1}{2}\angle AOB = \tfrac{1}{2} \times 2x = x$    (from Theorem 1, angle at centre)

Hence $\angle TAB = \angle ACB$, as required.

*Example*    A circle is drawn inside $\triangle ABC$, touching the sides $AB$, $BC$ and $CA$ at $L$, $M$ and $N$ respectively. If the angles at $A$, $B$ and $C$ are 66°, 68° and 46° respectively, find the angles of $\triangle LMN$.

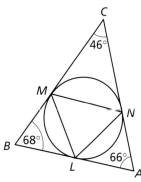

$\triangle ALN$ is isosceles (tangents from a point are equal). Hence $\angle ALN = \tfrac{1}{2}(180° - 66°) = 57°$.

$\angle LMN = \angle ALN = 57°$       (alternate segment theorem)

Similarly $\angle MNL = 56°$ and $\angle MLN = 67°$.
The angles of $\triangle LMN$ are 57°, 56° and 67°.

# 78 CIRCLES

## Exercise 5.7

You will need:
- compasses
- ruler
- protractor

1. Verify Theorem 5 by measurement. Draw a tangent $TA$ to a circle, draw a chord $AB$ and pick a point $C$ in the alternate segment. Verify that $\angle TAB = \angle ACB$.
2. Find the unknown angles in these diagrams.

   a
   
   b

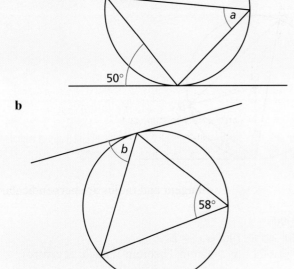

3. The angles of $\triangle ABC$ are $56°$, $54°$ and $70°$. A circle is drawn within the triangle, touching its sides at $X$, $Y$ and $Z$. Find the angles of $\triangle XYZ$.
4. The angles of $\triangle PQR$ are $82°$, $48°$ and $50°$. A circle is drawn within the triangle, touching its sides at $L$, $M$ and $N$. Find the angles of $\triangle LMN$.
5. The angles of $\triangle DEF$ are $61°$, $65°$ and $54°$. A circle is drawn through $D$, $E$ and $F$. The tangents to the circle at $D$, $E$, and $F$ form a triangle $ABC$. Find the angles of this triangle.

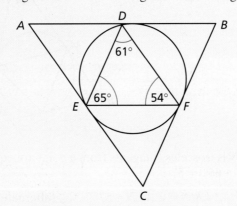

6. The angles of $\triangle PQR$ are $67°$, $52°$ and $61°$. A circle is drawn through $P$, $Q$ and $R$. Find the angles of the triangle formed by the tangents to the circle at $P$, $Q$ and $R$.

## Mixed examples

In the previous examples and exercises you have known which theorem to use. Sometimes you have to decide which theorem to use. In some problems you use more than one theorem.

*Example*  In the diagram, $AB$ is a diameter. $\angle CAB = 40°$. Find $\angle ADC$.

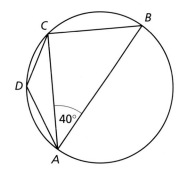

$\angle ACB = 90°$      (angle in a semicircle)
$\angle ABC = 50°$      (angle sum of a triangle)
$\angle ADC = 180° - 50°$   (angles of a cyclic quadrilateral)
$\angle ADC = 130°$.

## Exercise 5.8

**1** Find the unknown angles in these diagrams. The centre of each circle is labelled $O$.

**a**                **b**              **c**

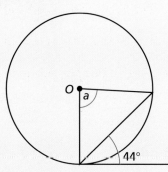 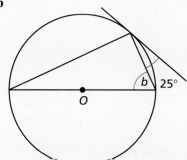 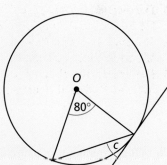

**2** This diagram is not drawn accurately.
  **a** If $a = 140°$, find $b$, $c$ and $d$.
  **b** If $b = 80°$, find $a$, $c$ and $d$.
  **c** If $c = 35°$, find $a$, $b$ and $d$.
  **d** If $d = 130°$, find $a$, $b$ and $c$.
  **e** If $a = d$, find $a$, $b$ and $c$.

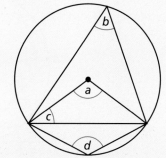

**3** In the diagram, $O$ is the centre of the circle. If $\triangle ACB = 136°$, find $\angle AOB$.

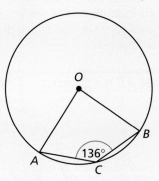

**4** In the diagram, $ABCD$ is a cyclic quadrilateral. $TA$ is the tangent to the circle at $A$. If $\angle TAC = 72°$, find $\angle ABC$.

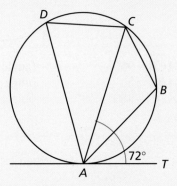

**5** In the diagram, $AB = AC$. Show that $\angle ADX = \angle ADB$.

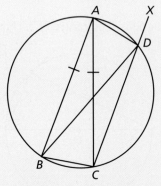

**6** $AB$ is a diameter of a circle, and $TC$ is a tangent. If $\angle TCA = 48°$, find $\angle BAC$.

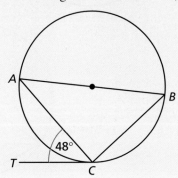

**7** In the diagram, *C* is the centre of the circle. If ∠*TAB* = 41°, find ∠*BCA*.

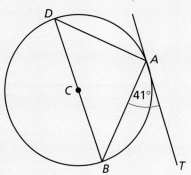

**8** In the diagram, *ABCD* is a parallelogram. *AB* cuts the circle at *E*. Show that triangle *CBE* is isosceles.

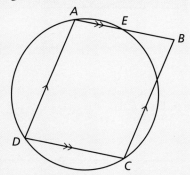

**9** In the diagram, *AB* is perpendicular to *CD*. The centre of the circle is *O*. If ∠*AOC* = 108°, find ∠*DOB*.

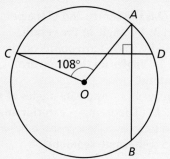

**10** In the diagram, *TA* is a tangent to the circle. *AB* and *CD* are parallel. *AC* and *BD* meet at *X*. ∠*BAT* = 63° and ∠*DCB* = 119°.

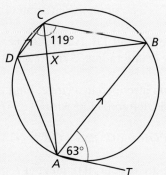

    **a** Find ∠*ADB*.
    **b** Find ∠*DAT*.
    **c** Show that △*XCD* is isosceles.

**11** In the diagram, $TCS$ is a tangent to the circle and $DA$ is a diameter. $AB = BC$ and $\angle DAC = 41°$.

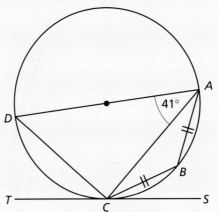

- **a** Find $\angle TCD$.
- **b** Find $\angle ACS$.
- **c** Find $\angle BCA$.
- **d** Find $\angle BCS$.
- **e** The radius of the circle is 10 cm. Find $CA$.

## SUMMARY

- Suppose $AB$ is a chord of a circle, $O$ is the centre and $C$ is a point on the circumference.
  Then $\angle AOB = 2 \times \angle ACB$. (Theorem 1 – **angle at the centre**)
- Suppose $AB$ is a chord of a circle and $C$ and $D$ are points on the circumference on the same arc.
  Then $\angle ACB = \angle ADB$. (Theorem 2 – **angles on the same arc**)
- Suppose $ABCD$ is a **cyclic quadrilateral**.
  Then $\angle ABC + \angle ADC = 180°$. (Theorem 3 – **opposite angles of a cyclic quadrilateral**). If $DC$ is extended to $E$, then $\angle DAB = \angle BCE$. (Theorem 4 – **exterior angle of a cyclic quadrilateral**)
- Suppose $AB$ is a chord, and $AT$ is a tangent. $C$ is a point in the alternate segment.
  Then $\angle TAB = \angle ACB$. (Theorem 5 – **alternate segment**)

## Exercise 5A

1. $AB$ is a diameter of a circle, and $C$ is a point on the circumference. If $AC = 7$ cm and $BC = 9$ cm, find $AB$.
2. The tangents from $T$ to a circle are $TA$ and $TB$. If $\angle ATB = 68°$, find $\angle TBA$.
3. A chord $AB$ is in a circle of centre $O$ and radius 23 cm. If $\angle OAB = 47°$, find the distance from $O$ to the chord.
4. $ABCDEFGHI$ is a regular nonagon inscribed in a circle. Find $\angle ABG$.
5. $A$, $B$, $C$ and $D$ are points on a circle, lettered clockwise. If $\angle ABD = 48°$ and $\angle ADB = 33°$, find $\angle BCD$.
6. $PQRS$ is a cyclic quadrilateral. $\angle PQR = 107°$ and $\angle QRS = 113°$. Find the other angles of the quadrilateral.
7. The chords $AB$ and $DC$ of a circle are produced to meet at $E$ outside the circle. $\angle DAB = 77°$ and $\angle BEC = 23°$. Find $\angle CBE$.
8. $PQRS$ is a cyclic quadrilateral. $TS$ is the tangent to the circle at $S$. If $\angle SRQ = 126°$, find $\angle QST$.

9 In the diagram, $X$ is the centre of the circle, and $TC$ is a tangent. If $\angle CXD = 84°$, find $\angle DCT$.

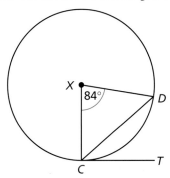

10 $ABC$ is an acute-angled triangle inscribed in a circle. Points $P$, $Q$ and $R$ are on the minor arcs of $AB$, $BC$ and $CA$ respectively. Show that
$$\angle APB + \angle BQC + \angle CRA = 360°$$

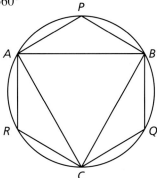

## Exercise 5B

1 $AB$ is a diameter of a circle, and $C$ is a point on the circumference. If $\angle ABC = 77°$, find $\angle BAC$.
2 A tangent from $T$ to a circle with centre $O$ is $TA$. If $TO = 53\,\text{cm}$ and $TA = 43\,\text{cm}$, find the radius of the circle.
3 A log of wood is a cylinder with radius $26\,\text{cm}$. A slice parallel to the axis of the cylinder is sawn off. If the greatest thickness of the slice is $3\,\text{cm}$, find the width of the slice.

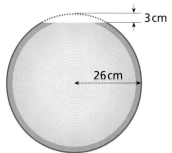

4 $AB$ is a diameter of a circle with centre $O$, and $C$ is a point on the circumference. If $\angle AOC = 88°$, find $\angle ABC$.
5 $P$, $Q$, $R$ and $S$ are points on a circle, lettered clockwise. $PR$ and $QS$ meet at $T$. $\angle TPQ = 35°$ and $\angle PTS = 68°$. Find $\angle PRS$.
6 $ABCD$ is a cyclic quadrilateral in a circle with centre $O$. If $\angle AOC = 104°$, find two possible values for $\angle ABC$.

7 A triangle has angles 82°, 56° and 42°. A circle is drawn touching the sides of the triangle. Find the angles of the triangle formed by the tangents at the three points of contact.

8 $PQ$ is a diameter of a circle, and $TR$ is a tangent, touching it at $R$. If $\angle RPQ = 43°$, find the two possible values for $\angle PRT$.

9 $ABCD$ is a cyclic quadrilateral with $AB = AD = DC$. $BC$ is produced to $E$, where $CE = AB$. Show that $ED$ is parallel to $CA$.

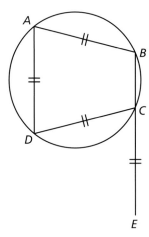

10 $ABCDEF$ is a cyclic hexagon (not necessarily regular). Show that

$$\angle A + \angle C + \angle E = 360°$$

## Exercise 5C

You will need:
- ruler
- compasses

This is a way to find square roots geometrically. It uses the theorem about the angle in a semicircle.

In the diagram, $AB$ is a diameter. We know that $\angle ACB = 90°$. The line $CD$ is the perpendicular from $C$ to $AB$.

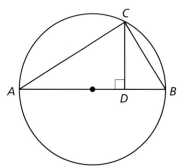

**1** Show that $\angle DCB = \angle DAC$ and $\angle CBD = \angle ACD$.

Clearly $\angle CDA = \angle BDC = 90°$. As $\triangle CDA$ has the same angles as $\triangle BDC$, the triangles are similar. Suppose the enlargement which takes $\triangle BDC$ to $\triangle CDA$ has scale factor $k$.

**2** Show that $\dfrac{CD}{BD} = \dfrac{DA}{DC} = k$.

**3** Show that $CD = \sqrt{BD \times DA}$.

So, if we can draw lines of length $x$ and 1 units, then we can draw a line of length $\sqrt{x}$ units.

**4** Follow this procedure to find $\sqrt{5}$: draw a line of length $5 + 1$, i.e. 6 inches, as shown. This is the diameter of the circle. Find the midpoint of the line, and draw the circle. Raise a perpendicular from $D$. This crosses the circle at $C$.

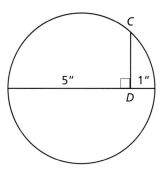

**5** Measure $DC$. How close is it to $\sqrt{5}$ inches?

**6** Use this method to find some other square roots.

# Exercise 5D

Start with any scalene triangle $ABC$ (so all the angles are different). The incircle touches the sides of the triangle at $P$, $Q$ and $R$. Find the angles of the triangle $PQR$ (as in the example on page 77). Repeat the process with the incircle of $\triangle PQR$. What happens as you continue?

# Exercise 5E

This is an exercise on constructions, exploiting some of the results of this chapter. Use straight edge and compasses only.

You will need:
- ruler
- compasses

**1** Mark two points $A$ and $B$. Construct the locus of points $P$ for which $\angle APB = 45°$.

**2** Mark two points $C$ and $D$. Construct the locus of points $Q$ for which $\angle CQD = 30°$.

**3** Draw an acute-angled triangle $XYZ$. Find the point $O$ for which

$$\angle XOY = \angle YOZ = \angle ZOX = 120°$$

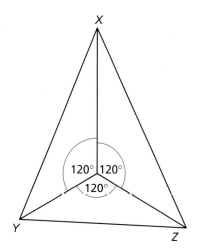

(This point is called the Euler centre of the triangle. It is the point in the triangle for which $OX + OY + OZ$ is as small as possible.)

> Leonhard Euler (1707–1783) was a Swiss mathematician and physicist who made great contributions to the subjects of geometry, calculus, mechanics and number theory. He lost the sight in one eye in 1735, then 30 years later became totally blind, but his extraordinary memory and aptitude for mental calculation meant that he could continue to work right up until his death in 1783.

# Proof

*'Mathematics is the only subject in which men have knowledge like that of gods.'*
Plato (427–347 BC)

> The most famous mathematics book of all time is the *Elements of Geometry*, written by Euclid, a Greek mathematician who lived about 300 BC. This book provided the standard of proof for over two thousand years.

Mathematics is different from all other subjects. What happened in history is often a matter of opinion. The value of a poem is a matter of personal taste. Even the results of science are found from observation. But the results of mathematics are *proved*. They are true, and no sensible person can argue with them.

Also, the results of mathematics are true everywhere and always. If a distant planet is found to have life, its biology will be completely different from ours. The universe could have developed differently, with different physics and chemistry. But on all planets and in all possible universes the results of mathematics are the same. It is always true, no matter how many tentacles you have, that $2 + 2 = 4$, and that the base angles of an isosceles triangle are equal.

In this book we provide proofs of many results, instead of just stating them. We will ask you to provide a proof of your own in some cases. First we need to make some distinctions about the way we use certain words in mathematics.

## Terminology

### Equation and identity

Consider the two expressions:

$$x^2 - 3x + 2 = 0 \qquad x^2 + 2x + 1 = (x+1)^2$$

The first expression is an equation. We can solve it, to find that $x = 1$ or $x = 2$. The equation is not true for $x = 3$, for example.

The second expression is an identity. It is true for all possible values of $x$. We cannot solve it, because it works for every value of $x$!

### Proof and verification

A proof of a result shows that it is true in all possible cases. In mathematics, a verification merely confirms that it is true in one particular case.

A verification of the identity $x^2 + 2x + 1 = (x+1)^2$ above is to put $x = 7$.

Left-hand side $= 7^2 + 2 \times 7 + 1 = 49 + 14 + 1 = 64$
Right-hand side $= (7+1)^2 = 8^2 = 64$

This confirms that the result is true for this particular case. It is not a *proof*. For a proof, we need to show that the left-hand side and the right-hand side are equal for *every* value of $x$. This can be done by expansion.

$$(x+1)^2 = (x+1)(x+1) = x^2 + 1 \times x + x \times 1 + 1^2 = x^2 + 2x + 1$$

So the left-hand side is identically equal to the right-hand side.

### Proof and counterexample

To show that a general statement is true you need to prove it. To show that it may be false you need to find a case in which it is false. This case is a **counterexample**.

Here are two examples: one where we prove that something is true, and another where we show that something is false.

To prove $(a+b)(a-b) = a^2 - b^2$
Take the left-hand side, and expand it.
$$(a+b)(a-b) = aa - ab + ba - bb$$
$$= a^2 - b^2 \quad (aa = a^2 \text{ and } ab = ba)$$

So the result is proved true for *all* values of $a$ and $b$.

This is a proof. When you prove something not true, that is a **disproof**.

*Suppose someone claims that $\sqrt{a+b} = \sqrt{a} + \sqrt{b}$. How can you show that it is false?*

We don't have to prove it false for every possible value of $a$ and $b$. Take any pair of values.

Take $a = b = 1$.

The left-hand side is $\sqrt{1+1} = \sqrt{2}$.
The right-hand side is $\sqrt{1} + \sqrt{1} = 1 + 1 = 2$.
These are different. So the suggested rule is false.

The pair of values, $a = 1$ and $b = 1$, is the counterexample.

Note the difference between the two examples. It may seem unfair – that to prove a general statement true you have to show it true in *all* cases, but that to show it false you only have to disprove it for *one* case. But it makes sense. If someone claims that 'All mathematicians are rich' then a proof would have to show that every mathematician in the world is rich. To disprove the statement, you only have to produce one poor mathematician.

# Exercise P1

1. Which of the following are equations and which are identities?
   a. $x^2 + 3x - 1 = x^2 - 3x + 1$
   b. $(a-b)^2 = a^2 - 2ab + b^2$
   c. $(x+3)^2 - x^2 = 3(2x+3)$
   d. $(x+1)^2 = x^2 + 1$
   e. $\sin x = \cos x$
   f. $\sin x = \cos(90° - x)$

2. Verify the identity $x^2 + 2x + 1 = (x+1)^2$ above by putting
   a. $x = 3$
   b. $x = -2$

3. Consider the equation $x^2 - x - 6 = 0$.
   a. Verify that it is true for $x = -2$ and $x = 3$.
   b. Find a value of $x$ for which the equation is not true.

4. Verify the identity $(a+b)(a-b) = a^2 - b^2$ by putting
   a. $a = 8$ and $b = 3$
   b. $a = 2$ and $b = 7$

5. Verify the identity $\dfrac{a}{a+b} = 1 - \dfrac{b}{a+b}$ ($a$ and $b$ positive) by putting
   a. $a = 5$ and $b = 4$
   b. $a = 7$ and $b = 3$

**88** • PROOF

**6** Construct a parallelogram. Measure its sides and angles. Verify that
  **a** opposite sides are equal
  **b** opposite angles are equal
  **c** the diagonals bisect each other.

You will need:
• ruler
• protractor

**7** Draw any triangle *ABC*. Measure its sides *a*, *b*, *c* and its angles *A*, *B*, *C*. Verify that the sine rule and cosine rule are true, i.e. that

$$\frac{\sin A}{a} = \frac{\sin B}{b} = \frac{\sin C}{c} \text{ and } a^2 = b^2 + c^2 - 2bc \cos A$$

**8** The following statements are all false. In each case provide a counterexample.
  **a** Every odd number is prime.
  **b** Every prime number is odd.
  **c** For all values of *a* and *b*, $(a+b)^2 = a^2 + b^2$.
  **d** For all positive values of *a*, *b* and *c*, $\frac{a}{b+c} = \frac{a}{b} + \frac{a}{c}$.
  **e** For all positive values of *a*, *b* and *c*, $\frac{a+b}{a+c} = \frac{b}{c}$.
  **f** If any two quadrilaterals have equal angles, then they are similar.
  **g** If $a > b$ and $c > d$, then $a - c > b - d$.
  **h** For all angles *a* and *b*, $\sin(a+b) = \sin a + \sin b$.

## Geometric proof

The geometric chapters of this book contain many examples of proof. See for example those in chapter 5, that the opposite angles of a cyclic quadrilateral add up to 180°, and so on.

You should know the conditions for two triangles to be congruent. They are

SSS  (corresponding sides in each triangle are equal)

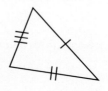 

SAS  (two sides and the included angle are equal in each triangle)

ASA  (two angles and the included side are equal in each triangle)

 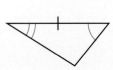

RHS  (both triangles are right-angled, the hypotenuse and one other side are equal)

Note that SSA (two sides and an angle not included) is not a condition for congruence.

In geometric proofs, you should always give reasons for each step.

If you are told beforehand that two lines are equal, write 'given' when you use this fact. For example

$$AB = CD \quad \text{(given)}$$

If a line or an angle is in two parts of the diagram, use the word 'common'. For example

$$AB \text{ is common to } \triangle ABC \text{ and } \triangle ABD$$

Below are two examples of geometric proof. Note the reasons for each step of the proof.

*Example*  $AXB$ and $CXD$ are straight lines meeting at $X$. $AX = CX$ and $BX = DX$. Show that $\triangle AXD$ and $\triangle CXB$ are congruent.

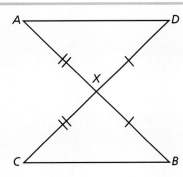

**Proof**
$$AX = CX \quad \text{(given)}$$
$$BX = DX \quad \text{(given)}$$
$$\angle AXD = \angle CXB \quad \text{(vertically opposite)}$$
Hence $\triangle AXD$ and $\triangle CXB$ are congruent   (SAS).

In all proofs, it should be clear where you are starting from. In the example below we prove a result about a parallelogram. We start from the definition of a parallelogram, that opposite sides are parallel. We do not assume the result that the opposite sides are equal, we *prove* it.

*Example*  $ABCD$ is a parallelogram. Show that $AB = CD$.

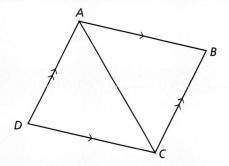

**Proof**
In the diagram, the parallelogram is $ABCD$. Join the diagonal $AC$.
$$\angle ACD = \angle CAB \quad \text{(alternate, as } AB \text{ and } CD \text{ are parallel)}$$
$$\angle BCA = \angle DAC \quad \text{(alternate, as } AD \text{ and } BC \text{ are parallel)}$$
$AC$ is common to both triangles.
   Hence $\triangle ACB$ and $\triangle CAD$ are congruent   (ASA).
Hence $AB = CD$.

## Exercise P2

**1** $AXB$ and $CXD$ are straight lines meeting at $X$. $AC$ and $DB$ are parallel, and $AX = XB$. Show that $\triangle AXC$ and $\triangle BXD$ are congruent.

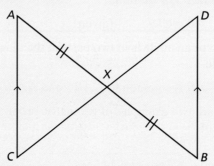

**2** $A$, $B$ and $C$ are points on a circle with centre $O$, with $AB = BC$. Show that $\triangle AOB$ and $\triangle COB$ are congruent.

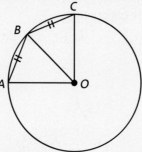

**3** $ABC$ is an equilateral triangle. $D$ and $E$ are points on $AB$ and $AC$ respectively, with $AD = CE$. Show that $\triangle ADC$ and $\triangle CEB$ are congruent.

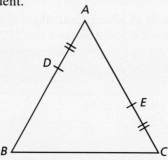

**4** $AB$ is a line segment, and $C$ is a point on it. $CD$ is perpendicular to $AB$, and $DA = DB$. Show that $\triangle DCA$ and $\triangle DCB$ are congruent.

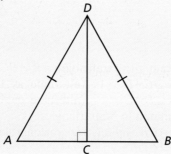

5 *ABCDEFGHI* is a regular nonagon (nine-sided polygon). Show that △*ABC* and △*BCD* are congruent.

6 *AB* and *CB* are lines meeting at *B*. *D* is a point on the bisector of ∠*ABC*, i.e. ∠*DBA* = ∠*DBC*. *E* and *F* are the feet of the perpendiculars from *D* to *AB* and *CB* respectively. Show that △*BDE* and △*BDF* are congruent.

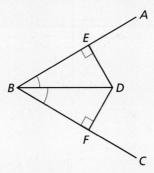

7 The diagram shows *ABCDE*, which is an equilateral triangle on top of a square. Show that △*EAB* and △*EDC* are congruent.

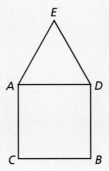

8 Let *ABC* be an isosceles triangle for which *AB* = *AC*. Let *M* be the midpoint of *BC*.

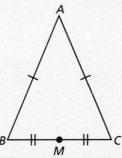

  **a** Show that △*AMB* and △*AMC* are congruent.
  **b** Show that ∠*ABC* = ∠*ACB*.
  **c** Show that ∠*AMB* = 90°.

**9** $ABCD$ is a kite, for which $AB = AD$ and $CB = CD$.
  **a** Show that $\triangle ACB$ and $\triangle ACD$ are congruent.

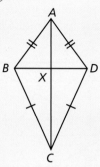

  The diagonals $AC$ and $BD$ meet at $X$.
  **b** Show that $\triangle AXB$ and $\triangle AXD$ are congruent.
  **c** Show that $AC$ and $BD$ are perpendicular.

**10** $ABCD$ is a parallelogram, i.e. $AB$ is parallel to $CD$ and $AD$ is parallel to $BC$. Extend the proof of the example on page 89 to show that $AD = BC$.

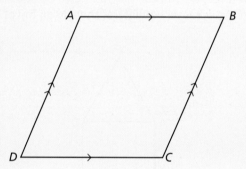

**11** $ABCD$ is a parallelogram.
  **a** Show that $\angle ABC = \angle ADC$ and $\angle BAC = \angle DCB$.
  **b** Let the diagonals meet at $X$. Show that the diagonals bisect each other, i.e. that $AX = CX$ and $BX = DX$.

Hint: use the result of the example above, that $AB = CD$.

**12** $ABCD$ is a quadrilateral in which $AB$ and $CD$ are equal and parallel. Show that $AD$ and $BC$ are also equal and parallel.

**13** $ABCD$ is a rectangle, i.e. all the angles are equal to 90°.

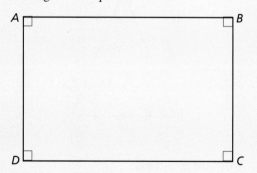

  **a** Show that opposite sides are parallel, i.e. that $AB$ is parallel to $CD$ and $AD$ is parallel to $BC$.
  **b** Show that opposite sides are equal, i.e. $AB = CD$ and $AD = BC$.
  **c** Show that the diagonals are equal, i.e. that $AC = BD$.

**14** *ABCD* is a rhombus, i.e. all the sides are equal.

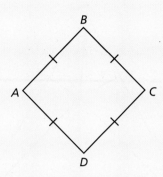

    **a** Show that opposite angles are equal.
    **b** Show that opposite sides are parallel.
    **c** Show that the diagonals bisect the angles.

**15** *ABCD* is a parallelogram. The bisectors of ∠*ABC* and ∠*DCB* meet at *X*. Show that ∠*BXC* = 90°.

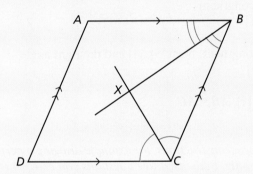

**16** *ABC* is a triangle. *X* is a point on *AB* for which *AX* = *BX* = *CX*. Show that ∠*ACB* = 90°.

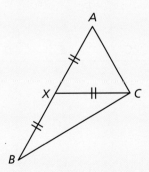

**17** *ABCD* is a trapezium, for which *AB* is parallel to *CD*. *AD* = *BC*. Is it always true that *AC* = *BD*?

18 What is wrong with the following 'proof' that every triangle is isosceles?

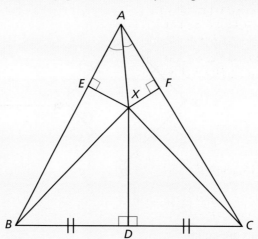

Let the bisector of $\angle BAC$ and the perpendicular bisector of $BC$ meet at $X$. Let the feet of the perpendiculars from $X$ to $BC$, $AB$ and $AC$ be $D$, $E$ and $F$ respectively. Join $XB$ and $XC$. Then

| | | |
|---|---|---|
| $\triangle AXE$ and $\triangle AXF$ are congruent | (ASA) | [1] |
| $\triangle XDB$ and $\triangle XDC$ are congruent | (SAS) | [2] |
| Hence $XE = XF$, from [1] | | [3] |
| $BX = CX$, from [2] | | [4] |
| $\triangle EXB$ and $\triangle FXC$ are congruent, from [3], [4] and the right angle | (RHS) | [5] |
| Hence $EB = FC$, from [5] | | [6] |
| But $AE = AF$, from [1] | | [7] |

Add the sides in [6] and [7]. $AB = AC$.
Hence $\triangle ABC$ is isosceles.

*You have met various geometrical constructions, for example drawing a perpendicular from a point to a line using ruler and compasses only. In the next five questions you prove the validity of these constructions.*

## 19 Drawing a perpendicular from a point to a line

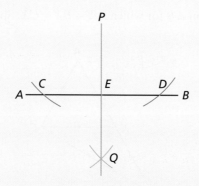

Let the point $P$ and the line $AB$ be as shown. Draw arcs from $P$, crossing $AB$ at $C$ and $D$. Draw arcs from $C$ and $D$, meeting at $Q$. $PQ$ crosses $AB$ at $E$. We want to prove that $PE$ is the perpendicular.
**a** Show that $\triangle PQC$ and $\triangle PQD$ are congruent.
**b** Show that $\angle CPE = \angle DPE$.
**c** Show that $\triangle CPE$ and $\triangle DPE$ are congruent.
**d** Show that $\angle PEC = 90°$.

**20 Drawing a perpendicular from a point on a line**
Let the point $P$ and the line $AB$ be as shown.
Draw arcs from $P$, crossing $AB$ at $C$ and $D$.
Draw arcs from $C$ and $D$, meeting at $Q$. We
want to prove that $PQ$ is the perpendicular.
  By considering $\triangle CQP$ and $\triangle DQP$, show
that $\angle QPC = 90°$.

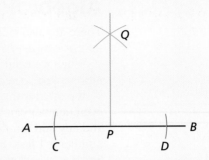

**21 Finding the perpendicular bisector of a line**
Let the line be $AB$ as shown. Draw arcs from
$A$ and $B$ meeting at $P$ and $Q$. Let $PQ$ cross
$AB$ at $E$. We want to show that $PE$ is the
perpendicular bisector.
  By considering $\triangle APQ$ and $\triangle BPQ$, and
then $\triangle AEP$ and $\triangle BEP$, show that $AE = BE$
and that $\angle AEP = 90°$.

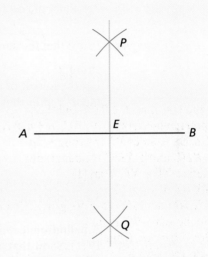

**22 Finding the bisector of an angle**
Let the angle be between the lines $l$ and $m$ as shown.
Draw arcs from $O$, meeting $l$ and $m$ at $A$ and $B$.
Draw arcs from $A$ and $B$, meeting at $C$. We want
to show that $OC$ is the bisector.
  Show that $\angle COA = \angle COB$.

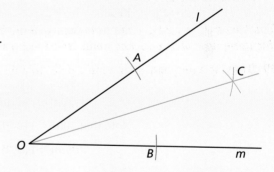

**23 Constructing an angle of 60°**
Draw an arc from $A$, cutting $AB$ at $C$.
Draw an arc from $C$, cutting the first arc at $D$.
Show that $\angle DAC = 60°$.

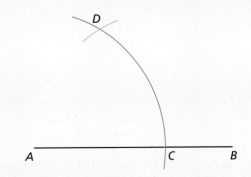

## Algebraic proof

There are also many examples of proofs which need algebra. Above we proved that $(a+b)(a-b) = a^2 - b^2$ by algebraically expanding the left-hand side.

When we want to prove something about whole numbers, we often let the number concerned be $n$. So the proof might begin: let the smaller number be $n$.

We often want to prove things about even or odd numbers. Note that any even number can be written as $2n$ (because it is twice another whole number) and any odd number can be written as either $2n - 1$ or as $2n + 1$ (because an odd number is 1 less than an even number, and also 1 greater than an even number.) So when we want to prove something about an odd number, the proof might begin: let the odd number be $2n - 1$.

*Examples*  Prove that the sum of three consecutive numbers is divisible by 3.

**Proof**
Let the lowest of the numbers be $n$. Then the next number is $n + 1$, and the number after that is $n + 2$. The sum of these three numbers is

$$n + n + 1 + n + 2$$
$$= 3n + 3$$
$$= 3(n + 1)$$

So 3 is a factor of the sum.
The sum of the three consecutive numbers is divisible by 3.

A **palindromic** number is a number which is the same when reversed (such as 1991). Show that any four-digit palindromic number is divisible by 11. (For example, $1991 = 11 \times 181$.)

**Proof**
Suppose the digits of the four-digit palindromic number are $m, n, n,$ and $m$. So the number is written as $mnnm$. (Here this does *not* mean $m \times n \times n \times m$.) This number represents

$$1000m + 100n + 10n + m$$
$$= 1001m + 110n$$

Note that $1001 = 11 \times 91$ and $110 = 11 \times 10$. Hence

$$mnnm = 11 \times 91m + 11 \times 10n = 11(91m + 10n)$$

Hence $mnnm$ is always divisible by 11.

# Exercise P3

In all these questions the numbers referred to are whole numbers.

1. Show that the sum of two consecutive numbers is odd.
2. Show that the sum of two consecutive odd numbers is divisible by 4.
3. Show that the sum of two consecutive even numbers is *never* divisible by 4.
4. Show that the sum of five consecutive numbers is divisible by 5.
5. Show that the sum of seven consecutive numbers is divisible by 7.
6. You are probably familiar with number games which begin: 'Think of a number...'.
   Now you can prove why they work.
   
   **a** 'Think of a number. Add 5, double, take away 10 and then halve. You are back with the number you started with.' Let the first number be $n$. Show that the final number is also $n$.
   
   **b** 'Think of a number, double it, add 34, halve, take away the number you first thought of. You are left with 17.' Prove this.

7. Show that the product of two consecutive numbers is even.
8. Show that the product of three consecutive numbers is divisible by 6.
9. Show that the product of four consecutive numbers is divisible by 24.
10. Prove that, if $n$ is even, $n^2$ is divisible by 4.
11. Look at the following square numbers.

    $$4 = 2^2 \quad 9 = 3^2 \quad 16 = 4^2 \quad 25 = 5^2$$
    
    **a** In each case above, find the remainder after division by 4. You should find that it is either 0 or 1. Verify this for some other square numbers.
    
    **b** Prove that, for any square number, the remainder on division by 4 is either 0 or 1. (Hint: the original number $n$ is either even or odd. So it is either $2r$ or $2r + 1$, for some $r$).

12. Take any two digit number $mn$. Show that, if the sum $m + n$ of its digits is divisible by 3, then the original number is divisible by 3. (Hint: consider the difference between $mn$ and $m + n$).
13. Take any numbers $a$ and $b$ with $a > b$. Show that $2ab$, $a^2 - b^2$ and $a^2 + b^2$ form a Pythagorean triad (i.e. they are the lengths of a right-angled triangle).
14. Show that any six-digit palindromic number is divisible by 11.
15. Show that not every three-digit palindromic number is divisible by 11.
16. Can you make a general statement, from the example above about four-digit palindromic numbers and questions 14 and 15?
17. Show that the sum of two odd squares cannot be a square.

# 6 Statistical diagrams

As part of the millennium events of 2000, a village made a survey of the ages of its houses. The results are in the table below.

| date built | number of houses |
|---|---|
| 1501–1800 | 23 |
| 1801–1900 | 48 |
| 1901–1950 | 53 |
| 1951–1960 | 27 |
| 1961–1970 | 38 |
| 1971–1980 | 33 |
| 1981–1990 | 29 |
| 1991–2000 | 39 |

The results were displayed in a bar chart in the parish magazine.

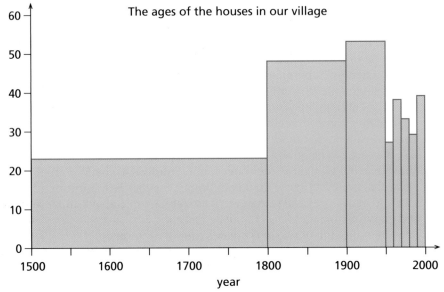

Notice that the biggest block is for the houses in the 1501–1800 range. It looks as though the majority of the village's houses are more than two centuries old! This is extremely misleading, as only about 8% of the houses are in that range. Similarly, the houses built in the last decade of the millennium are only represented by a very thin bar, although more house building was going on then than during any other decade.

The difficulty arose because of the widths of the intervals. The longest interval had width 300 years, 30 times more than that of the more recent intervals. We need a way to draw a diagram which takes account of the widths of the intervals. The most appropriate diagram for this is a **histogram**.

# Histograms

When you look at a statistical diagram, it should be the *area* of each region which tells you how much data it represents. So, in a histogram, we want the areas of the rectangles to be proportional to the numbers of data in the intervals. We can do this by deciding upon a standard width of interval, and then adjusting the heights of the bars which don't have standard widths. If an interval is twice the standard width then halve its height, and if an interval is a third the standard width then triple its height.

Here is another way of looking at it. Suppose the standard width is 10 units. Then an interval of width 20 units contains two standard intervals. If the data is evenly split between these two intervals then the bars are half the height of the frequency.

Suppose the standard width is 10 units. We label the vertical axis as 'frequency per 10 unit interval'. The vertical axis measures the **frequency density**, i.e. the frequency per given interval. (We can also label the vertical axis 'frequency density'.)

*Example*  A firm conducted a survey into the earnings of its employees. The results are below. Construct a histogram to show the data.

| salary, $x$ (£1000s) | frequency |
|---|---|
| $10 \leq x < 15$ | 11 |
| $15 \leq x < 20$ | 34 |
| $20 \leq x < 30$ | 42 |
| $30 \leq x < 40$ | 12 |
| $40 \leq x < 50$ | 7 |
| $50 \leq x < 100$ | 10 |

Notice that the intervals are not equal. Take a width of £10 000 as the standard. Then the first two intervals have half this width, and the sixth interval has 5 times the width. Adjust by doubling the frequencies in the first two intervals and dividing the frequency in the sixth interval by 5. The frequencies are converted to frequency densities. The table now becomes:

| salary, $x$ (£1000s) | frequency per £10 000 interval |
|---|---|
| $10 \leq x < 15$ | 22 |
| $15 \leq x < 20$ | 68 |
| $20 \leq x < 30$ | 42 |
| $30 \leq x < 40$ | 12 |
| $40 \leq x < 50$ | 7 |
| $50 \leq x < 100$ | 2 |

Now draw the histogram. Label the horizontal axis as 'salary', and the vertical axis as 'frequency per £10 000 interval'. The result is shown.

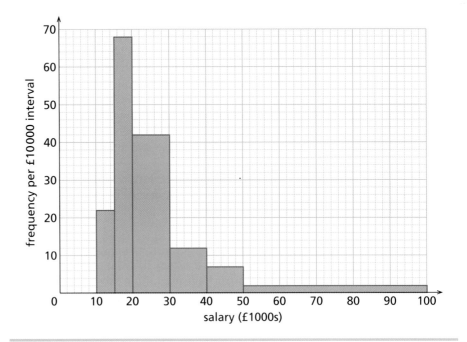

**Note.** If we had not adjusted the heights to take account of the different widths, the result would be as shown below. Notice that there seem to be far too many people in the last interval and too few in the first.

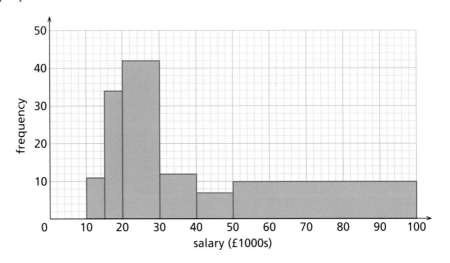

## Exercise 6.1

**You will need:**
- graph paper

**1** A survey is done into the departure time of trains. The table below gives the results. Draw a histogram to display the data.

| time late, $t$ (minutes) | frequency |
|---|---|
| $0 \leq t < 0.5$ | 13 |
| $0.5 \leq t < 1$ | 10 |
| $1 \leq t < 2$ | 18 |
| $2 \leq t < 3$ | 17 |
| $3 \leq t < 4$ | 10 |
| $4 \leq t < 5$ | 5 |
| $5 \leq t < 15$ | 25 |

**2** 100 children went on a cross-country run. Their times, in minutes, are given below. Draw a histogram to display the data.

| time | 15– | 20– | 25– | 30–40 |
|---|---|---|---|---|
| frequency | 38 | 21 | 18 | 23 |

**Remember:** The interval 15– means any time from 15 minutes up to 20 minutes.

**3** The marks obtained by 200 candidates for an exam are as follows. Draw a histogram to display the data.

| mark | 0–19 | 20–39 | 40–49 | 50–59 | 60–100 |
|---|---|---|---|---|---|
| frequency | 23 | 52 | 37 | 38 | 50 |

**4** The length of service of the employees of a company was found. The table below gives the data. Draw a histogram to display the data.

| length, $t$ (years) | $0 \leq t < 1$ | $1 \leq t < 2$ | $2 \leq t < 4$ | $4 \leq t < 6$ | $6 \leq t < 8$ | $8 \leq t < 10$ | $10 \leq t < 20$ |
|---|---|---|---|---|---|---|---|
| frequency | 21 | 14 | 23 | 16 | 13 | 10 | 21 |

**5** The midday temperature was measured for 60 days. The results are given in the table below. Draw a histogram to display the data.

| temperature, $t$ (°C) | $15 \leq t < 20$ | $20 \leq t < 25$ | $25 \leq t < 30$ | $30 \leq t < 40$ |
|---|---|---|---|---|
| frequency | 8 | 25 | 17 | 10 |

**6** At the beginning of this chapter there was a table giving the ages of the houses in a village.
 **a** Draw a histogram to display the data.
 **b** Compare your histogram with the bar chart produced in the parish magazine. Explain why the histogram gives a better picture than the bar chart.

**7** Draw a histogram for the data in question 1, but without making allowance for the different widths of intervals. Explain why the diagram is misleading.

## 102 STATISTICAL DIAGRAMS

From a histogram we can reconstruct the frequency table from which it came. The standard width of interval is given on the vertical axis. If an interval is half the standard width, then its frequency will have been doubled to find the height of the bar. So *halve* the height of the bar to find the original frequency. Similarly, if an interval is twice the standard width, then its frequency will have been halved to find the height of the bar. So *double* the height of the bar to find the original frequency.

From a histogram or a frequency table you can estimate the number of data within a certain interval.

*Examples* The histogram shows the times taken by runners to finish a race.

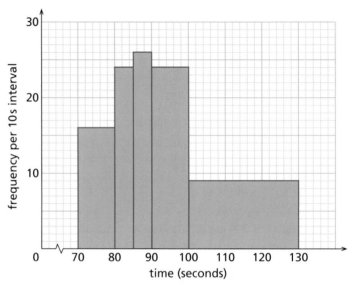

**a** Complete a frequency table for the data.
**b** Estimate the number of runners who took longer than 95 seconds.

**a** The standard width of interval is 10 seconds. The relative frequencies, i.e. the heights of the bars, are 16, 24, 26, 24 and 9.

The first interval is of width 10 seconds, and so it does contain 16 runners.

The next two intervals are of width 5 seconds, and so their frequencies must have been doubled. These two intervals contain $24 \div 2$ and $26 \div 2$ runners respectively.

The next interval is of standard width 10 seconds, and so it does contain 24 runners.

The last interval is of width 30 seconds, and so its frequency must have been divided by 3. It contains $9 \times 3$ runners.

The frequency table is as shown.

| time, $t$ (seconds) | $70 \leq t < 80$ | $80 \leq t < 85$ | $85 \leq t < 90$ | $90 \leq t < 100$ | $100 \leq t < 130$ |
|---|---|---|---|---|---|
| frequency | 16 | 12 | 13 | 24 | 27 |

**b** Assume that half the runners in the 90 to 100 interval took longer than 95 seconds. Add these to the runners in the 100 to 130 interval.

$$\tfrac{1}{2} \times 24 + 27 = 39$$

Estimate that 39 runners took longer than 95 seconds.

## Exercise 6.2

**1** The marks obtained in an exam are given in the histogram shown. Copy and complete the frequency table below.

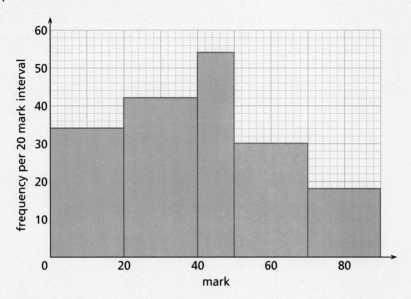

| mark | 0–19 | 20–39 | 40–49 | 50–69 | 70–89 |
|---|---|---|---|---|---|
| frequency | | | | | |

**2** A coach journey is advertised as taking 90 minutes. The histogram shown gives the journey times of the coach, measured over a period. Copy and complete the frequency table.

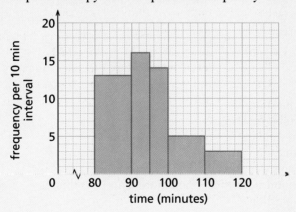

| time, $t$ (minutes) | $80 \leq t < 90$ | $90 \leq t < 95$ | $95 \leq t < 100$ | $100 \leq t < 110$ | $110 \leq t < 120$ |
|---|---|---|---|---|---|
| frequency | | | | | |

3 The amounts spent by a group of tourists are shown in this histogram. Construct a frequency table for the data.

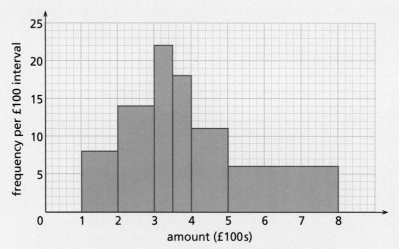

4 The temperature in a resort was measured over a period in the summer. This histogram shows the results. Construct a frequency table for the data.

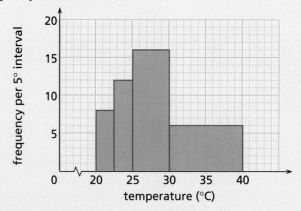

5 Refer to the situation in question 1.
  a How many candidates took the exam?
  b Estimate the number of candidates who scored less than 45.
6 Refer to the situation in question 2.
  a How many coach journeys were there?
  b Estimate the number of journeys which took more than 105 minutes.
7 Refer to the situation in question 3.
  a How many tourists were there?
  b Estimate the number of tourists who spent more than £600.
8 Refer to the situation in question 4. Estimate the number of days on which the temperature was less than 29 °C.

# Box plots

Suppose that data are presented in a frequency table. If we record a running total of the frequencies, the result gives the **cumulative frequencies**. If we plot the cumulative frequencies against the upper limits of the intervals, the result is a **cumulative frequency curve** or **ogive**.

A point on a frequency curve tells us how many items were less than a certain value. If $(x, y)$ is on the curve, then $y$ of the items had value less than $x$.

Look at this cumulative frequency curve. From it we can find the **median** and the **quartiles**.

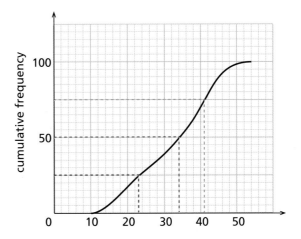

The median is the value that cuts off half of the frequency. In the diagram it is shown in red. The median is 34.

The lower quartile is the value that cuts off the bottom quarter of the frequency. In the diagram it is shown in blue. The lower quartile is 23.

The upper quartile is the value that cuts off the top quarter of the frequency. In the diagram it is shown in green. The upper quartile is 41.

These features of a set of data can be shown on a diagram called a **box plot** (also called a **box-and-whisker diagram**). They are especially useful for comparing sets of data.

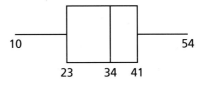

The box contains the central 50% of the data, with the lower quartile at the left and the upper quartile at the right. The position of the median is also shown.

The whiskers stretch from the ends to the limits of the data. So at the right side there is a whisker stretching from the upper quartile to the largest value, and at the left the whisker stretches from the lower quartile to the least value.

So a box plot normally shows five facts about the data: the minimum, the lower quartile, the median, the upper quartile and the maximum.

> These facts are sometimes called the five **indicators**.

# 106  STATISTICAL DIAGRAMS

The box plot below shows data between 32 and 58, with 40 as the lower quartile, 45 as the median and 49 as the upper quartile.

*Examples*  A box-and-whisker diagram is shown. Find the range and the interquartile range.

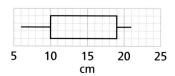

The ends of the whiskers give the least and greatest values. These are 6 cm and 21 cm.
The range is 15 cm.
   The ends of the box give the lower and upper quartiles. These are 10 and 19.
The interquartile range is 9 cm.

Construct a box-and-whisker diagram for which the values go between 45 seconds and 92 seconds, the quartiles are 62 seconds and 85 seconds, and the median is 70 seconds.

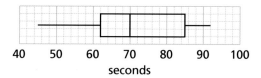

Draw the box between 62 and 85. Mark in the median at 70. Stretch the whiskers up to 92 and down to 45. The result is shown.

## Exercise 6.3

**You will need:**
- graph paper

**1** The diagram shows a cumulative frequency curve for exam marks, which go from 20 to 90. Find the median and the quartiles, and hence sketch a box plot for the data.

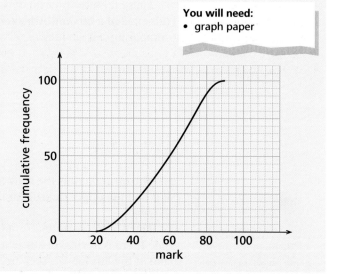

2 The diagram shows a cumulative frequency curve for the distances driven in a week by 200 motorists. Find the median and the quartiles, and hence sketch a box plot for the data.

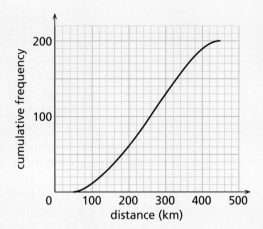

3 A box plot for the times of a race is shown. Find the range and the interquartile range.

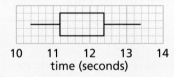

4 A box plot for the midday temperatures over 60 days is shown. Find the range and the interquartile range.

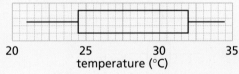

5 The spending money for 100 holidaymakers ranged from £80 to £230. The lower quartile, median and upper quartile were £130, £160 and £200 respectively. Sketch a box plot for the data.
6 A group of 80 users of the internet were asked how long they had spent on-line in the last week. The answers ranged from 2 hours to 20 hours, and the lower quartile, median and upper quartile were 4 hours, 6 hours and 10 hours respectively. Sketch a box plot for the data.
7 Sketch a cumulative frequency curve for the data in question 5.
8 Sketch a cumulative frequency curve for the data in question 6.

### Comparison

When two box plots are put alongside each other we can make immediate comparisons between the data they represent, provided that the plots are drawn to the same scale. Things to note are:

1 size – if one diagram is largely to the right of the other, then it represents larger values.
2 spread – if the box and the whiskers are wider in one diagram than in the other then it contains a wider range of values.

## 108 STATISTICAL DIAGRAMS

*Example* The same exam was taken by two groups of students (A and B). The box-and-whisker diagrams below show the results. Compare the groups.

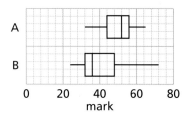

Notice that, for group A, the minimum, quartiles and the median are all greater than for group B. The maximum is greater for group B, but that may have been due to a single exceptional student.

On the whole, the marks for group A are higher than for group B.

Notice that, for group A, the range is less than for B, and that the interquartile range (the width of the box) is also less.

There is a narrower spread of results in group A.

## Exercise 6.4

**You will need:**
- graph paper

**1** The diagram shows box plots for the midday temperatures in two places. Comment on the differences.

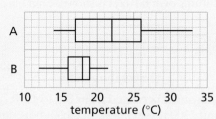

**2** An exam consisted of two papers. The diagram shows box plots for the results. Comment on the differences.

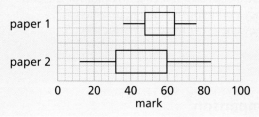

**3** Draw box plots for the following. Comment on the differences.

| group | minimum | lower quartile | median | upper quartile | maximum |
|---|---|---|---|---|---|
| A | 30 | 45 | 50 | 58 | 60 |
| B | 36 | 49 | 52 | 62 | 65 |

**4** Data showing the heights of plants are illustrated in the box plot. A sample of the same species of plants was grown with an added fertiliser. The minimum, lower quartile, median, upper quartile and maximum heights for the plants with fertiliser were 23 cm, 31 cm, 37 cm, 42 cm and 48 cm respectively. Draw a box plot for these plants below a copy of the first box plot, and comment on the differences.

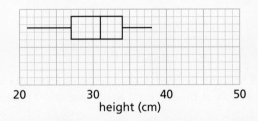

**5** The diagram shows cumulative frequency curves for the time in hours spent by girls and boys revising for GCSE maths. Draw box plots for the data, and comment on the differences.

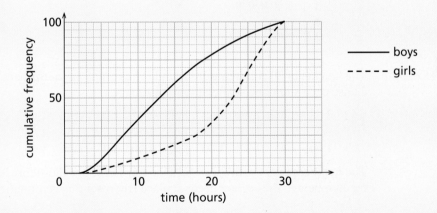

# Stem-and-leaf diagrams

> Both box plots and stem-and-leaf diagrams were invented by an American mathematician, John Tukey (1915–2000), as a means to display statistical data. His interest in statistics stemmed from a period during the Second World War when he worked for the US Fire Control Research Office, whose studies of weapons involved statistical problems.

A **stem-and-leaf diagram** is a sort of bar chart, in which the bars are made from the data themselves. It has the advantage that all the data can be recovered from the diagram (unlike a histogram, cumulative frequency curve or box plot).

Suppose the data consist of numbers in the twenties, thirties and forties, such as the following.

```
20  21  24  24  26  27  28  29  29
31  33  33  35  38  38  39  39
40  40  42  43  47  49
```

The stems will be the numbers 2, 3 and 4. The leaves will be the digits, arranged in increasing order. The digits should be aligned in columns. The diagram is as follows.

| 2 | 0 1 4 4 6 7 8 9 9 |
| 3 | 1 3 3 5 8 8 9 9 |
| 4 | 0 0 2 3 7 9 |

2 | 3 means 23

The same diagram would be used if the data were 200, 210, 240, etc. So we need the key at the top right to explain that 2|3 means 23, not 230, or 2300, or 0.23.

**Note.** The data were split into stems each representing 10 units. Each of these stems could be split into two, representing 5 units. The result would be

| 2 | 0 1 4 4 |
| 2 | 6 7 8 9 9 |
| 3 | 1 3 3 |
| 3 | 5 8 8 9 9 |
| 4 | 0 0 2 3 |
| 4 | 7 9 |

2 | 3 means 23

*Example* The heights of 30 children were found. The results, in metres, are shown below. Construct a stem-and-leaf diagram for the data.

```
1.63  1.58  1.54  1.66  1.68  1.50  1.52  1.48  1.40  1.47
1.41  1.52  1.39  1.38  1.47  1.66  1.50  1.37  1.44  1.55
1.63  1.60  1.45  1.37  1.40  1.51  1.60  1.47  1.66  1.66
```

The numbers begin with 1.3, 1.4, 1.5 or 1.6. These are the stems. The leaves are the numbers in the second decimal place (representing centimetres).

The stem-and-leaf diagram is below.

| 13 | 7 7 8 9 |
| 14 | 0 0 1 4 5 7 7 7 8 |
| 15 | 0 0 1 2 2 4 5 8 |
| 16 | 0 0 3 3 3 6 6 6 6 8 |

13 | 2 means 1.32 metres

# Exercise 6.5

1. The times, in seconds, of a group of runners to complete a race are shown. Draw a stem-and-leaf diagram to show the data, using stems of 40, 50 and 60, each containing 10 possible values.

   | | | | | | | | | | |
   |---|---|---|---|---|---|---|---|---|---|
   | 48 | 52 | 57 | 49 | 50 | 58 | 63 | 61 | 48 | 60 |
   | 49 | 60 | 63 | 68 | 66 | 55 | 58 | 53 | 52 | 60 |
   | 47 | 62 | 54 | 59 | 65 | 49 | 58 | 52 | 50 | 63 |

2. Draw a stem-and-leaf diagram for the data in question 1 in which each stem covers 5 possible values.

3. Forty people were asked to draw, without a ruler, two points 10 cm apart. The distances between the points were then measured, and the results are below. Construct a stem-and-leaf diagram to show the data.

   | | | | | | | | | | |
   |---|---|---|---|---|---|---|---|---|---|
   | 8.4 | 9.4 | 10.2 | 11.3 | 11.0 | 12.4 | 10.8 | 9.7 | 9.9 | 8.1 |
   | 7.3 | 10.2 | 11.2 | 12.5 | 10.3 | 9.3 | 10.4 | 11.4 | 8.4 | 8.2 |
   | 7.6 | 11.4 | 10.3 | 12.4 | 10.0 | 9.5 | 8.3 | 9.0 | 11.0 | 11.1 |
   | 8.8 | 10.7 | 7.9 | 10.8 | 12.4 | 9.4 | 12.2 | 8.0 | 9.2 | 9.5 |

## Back-to-back diagrams

A stem-and-leaf diagram can be used to compare sets of data. It is usual to put the leaves 'back-to-back' on either side of the stems.

*Example* A group of 20 students ran 100 m, before and after lunch. The results are below. Draw a back-to-back stem-and-leaf diagram and comment on what you notice.

**Before lunch**

| | | | | | | | | | |
|---|---|---|---|---|---|---|---|---|---|
| 12.8 | 13.3 | 12.9 | 12.6 | 13.9 | 14.6 | 14.1 | 12.5 | 13.4 | 14.0 |
| 13.8 | 13.0 | 14.7 | 14.8 | 13.2 | 14.6 | 14.2 | 12.7 | 13.0 | 14.4 |

**After lunch**

| | | | | | | | | | |
|---|---|---|---|---|---|---|---|---|---|
| 13.9 | 14.9 | 13.9 | 13.8 | 15.2 | 14.9 | 15.6 | 14.1 | 14.8 | 16.3 |
| 14.7 | 14.9 | 16.0 | 16.2 | 14.9 | 16.9 | 17.3 | 17.1 | 15.1 | 16.0 |

Have stems of 12, 13 and so on, up to 17. Put the times before lunch to the left of the stems, and the times after lunch to the right. The result is shown.

```
        9 8 7 6 5 | 12 |                    12 | 2 means 12.2
    9 8 4 3 2 0 0 | 13 | 8 9 9
  8 7 6 6 4 2 1 0 | 14 | 1 7 8 9 9 9 9
                  | 15 | 1 2 6
                  | 16 | 0 0 2 3 9
                  | 17 | 1 3
```

Notice that the after-lunch times are greater, and more spread out.
The students were slower after lunch, and there was a greater spread of time.

## Exercise 6.6

1. In question 3 of exercise 6.5, forty people were asked to draw, without a ruler, two points 10 cm apart. They are now told the distance apart of the points they had drawn, and are given a second chance. The new results are below. Show the results on a back-to-back stem-and-leaf diagram with the previous results. What do you notice?

   9.4  10.8  10.1  10.7  11.3  10.2  9.7  10.3  10.2  12.1
   10.4 10.0  10.5  11.3  10.7  10.2  10.1 10.4  9.0   10.9
   11.1 10.1  10.0  11.4  10.1  9.8   9.6  9.7   9.8   10.2
   9.8  10.1  12.0  10.2  11.8  10.3  11.1 11.0  10.5  9.9

2. Twenty children did the high jump. They were then given training, and given another go. The results, in metres, are below. Draw a stem-and-leaf diagram (with stems for each 0.5 m) and comment.

   **Before training**
   1.3  1.5  0.8  0.5  0.9  1.1  1.4  1.1  1.0  0.9
   1.3  1.2  1.1  0.8  0.9  1.3  1.5  1.6  1.4  0.6

   **After training**
   1.6  1.7  1.2  0.9  1.4  1.6  1.7  1.5  1.3  1.2
   1.7  1.6  1.2  1.4  1.0  1.5  1.7  1.7  1.5  0.7

3. The prices of takeaway meals bought by 20 customers in each of two restaurants were found. The results are below. Draw a back-to-back stem-and-leaf diagram, and comment.

   **Humayun's Tomb Curry Restaurant**
   5.3  6.8  6.5  7.3  8.1  5.0  4.7  5.9  8.3  7.7
   4.1  5.0  6.3  6.0  8.9  7.5  7.0  4.7  8.1  6.6

   **Catacombs Pizza Parlour**
   3.9  3.8  4.1  4.8  3.9  4.0  5.1  3.6  4.0  5.2
   4.7  3.5  4.8  5.7  5.0  4.3  5.9  4.4  3.7  5.0

## Summary

# SUMMARY

- In a **histogram**, the frequency in each interval is indicated by the area of its rectangle. Decide upon a standard width of interval, then adjust the height of the bars of those intervals which do not have the standard width. The height gives the **frequency density** in each interval.
- From a histogram, the original frequencies can be found.
- In a **box plot** (or **box-and-whisker diagram**), there is a central box that stretches between the lower and upper quartiles. There are whiskers that stretch from the lower quartile to the minimum value and from the upper quartile to the maximum value.
- Box plots are very useful for comparing two sets of data. In particular, they can compare the size and the spread of the data.
- A **stem-and-leaf diagram** is a bar chart in which the bars consist of the actual data. All the data can be recovered from the diagram.
- Back-to-back stem-and-leaf diagrams enable two sets of data to be compared.

## Exercise 6A

You will need:
- graph paper

**1** The table below gives the lengths of time that 120 internet users spent on-line in a given week. Construct a histogram to show the data.

| time, $t$ (hours) | $0 \leq t < 0.5$ | $0.5 \leq t < 1$ | $1 \leq t < 2$ | $2 \leq t < 3$ | $3 \leq t < 4$ | $4 \leq t < 10$ | $10 \leq t < 20$ |
|---|---|---|---|---|---|---|---|
| frequency | 23 | 17 | 15 | 14 | 10 | 19 | 22 |

**2** The histogram shown gives the numbers of goals scored by various teams in a season.

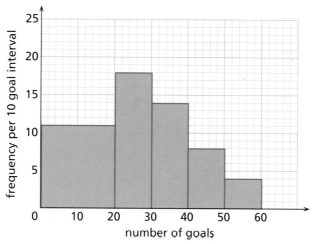

Copy and complete the frequency table below.

| number of goals | 0–19 | 20–29 | 30–39 | 40–49 | 50–59 |
|---|---|---|---|---|---|
| frequency | | | | | |

**3** Refer to question 2. Find the total number of teams shown.

**4** Refer to question 2. Estimate the proportion of teams who scored at least 45 goals.

## 114 STATISTICAL DIAGRAMS

**5** For the box plot below, find the range and the interquartile range.

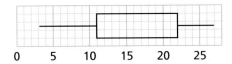

**6** The diagram shows a cumulative frequency curve for the lengths of time that people managed to hold their breath. Find the median and quartiles, and hence draw a box plot for the data.

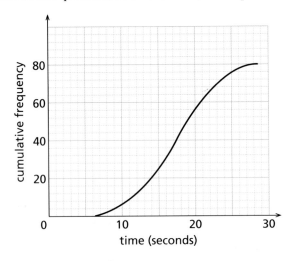

**7** The data below relate to the midday temperatures (in °C) in two places over the year. Draw box plots for both places, and comment.

| place | minimum | lower quartile | median | upper quartile | maximum |
|---|---|---|---|---|---|
| Bognor Regis | 3 | 12 | 17 | 23 | 29 |
| Ulan Bator | −25 | 10 | 23 | 27 | 38 |

**8** The heights (in cm) of 20 people are given below. Draw a stem-and-leaf diagram to illustrate the data.

160  173  188  166  184  190  188  165  177  159
168  166  173  194  188  192  180  174  172  183

**9** The numbers below are the marks obtained by two classes in a test. Draw a back-to-back stem-and-leaf diagram, with a stem for every five marks, and comment.

**Class A**
34  38  27  26  29  31  38  27  28  22
33  30  38  31  27  26  32  38  39  26

**Class B**
17  24  36  19  20  35  22  15  18  26
38  37  15  16  22  25  30  28  21  39

# Exercise 6B

**You will need:**
- graph paper

**1** The table below gives the prices of second-hand cars advertised in a local newspaper. Plot a histogram to show the data.

| price (£) | 0–499 | 500–999 | 1000–1999 | 2000–2999 | 3000–3999 | 4000–8000 |
|---|---|---|---|---|---|---|
| frequency | 12 | 15 | 17 | 15 | 12 | 24 |

**2** This histogram gives the lengths of the feature films shown in a cinema. Construct a frequency table for the data.

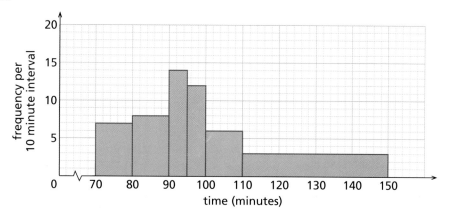

**3** Refer to question 2. Estimate the number of films which were longer than 105 minutes.

**4** This diagram is a box plot for the weights of 80 parcels. Draw a cumulative frequency graph to show the same information.

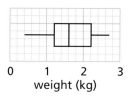

**5** A factory has two machines that produce the same type of component. The box plots below give the diameters of the components produced by each machine. Comment on the difference. Which is the better machine?

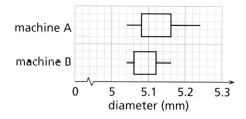

**6** The salaries of the office staff of a company range from £12 000 to £35 000, and the lower quartile, the median and the upper quartile are £16 000, £22 000 and £25 000 respectively. Draw a box plot to show this information.

**7** The distances, in miles, that 30 children travel to school are given below. Construct a stem-and-leaf diagram to show the data.

| 0.5 | 1.1 | 1.0 | 2.6 | 1.8 | 3.0 | 1.7 | 0.9 | 0.2 | 3.4 |
| 1.3 | 2.0 | 3.0 | 3.8 | 2.6 | 1.2 | 0.2 | 2.9 | 2.3 | 2.1 |
| 0.6 | 1.0 | 2.1 | 3.1 | 3.3 | 2.3 | 2.2 | 1.9 | 1.8 | 2.4 |

**8** The numbers below give the attendances, in 1000s, of matches played at two football grounds. The first ground belongs to a team near the top of the division, the second belongs to a team in the middle. Draw a back-to-back stem-and-leaf diagram and comment on the data.

**First ground**
76  68  71  75  78  76  70  77  74  75
77  68  70  71  73  67  78  77  72  79

**Second ground**
54  62  48  72  66  60  77  43  51  50
59  50  63  76  49  53  60  55  62  72

## Exercise 6C  

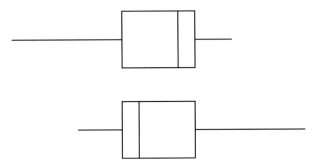

In the box plots of this chapter we have concentrated on two things: the magnitude of the data and the spread. These are not the only qualities that can be illustrated by a box plot.

Look at the two box plots on the right. What is the difference between them? What sort of data could have box plots like these? Think of situations which give rise to data like these.

## Exercise 6D

In this chapter all the data were provided for you. You can collect your own data and illustrate them on one of the sorts of diagram of this chapter. One suggestion is:

- Ask a sample of people to say when they think a minute has passed. Record the real time that has elapsed. Display the results on a stem-and-leaf diagram. Do people overestimate or underestimate?
  Tell the people sampled how much they have overestimated or underestimated, then give them a second go. Display the results back-to-back on the first results.
  Have people overcompensated? If they were told that their minute was too short the first time, is their next minute still too short or is it too long?

## Exercise 6E

**You will need:**
- computer with spreadsheet or other diagram-drawing package installed

There are many computer packages which can plot the statistical diagrams of this chapter. In particular, a spreadsheet may have the facility to produce diagrams. If you have access to such a package, use it to draw some of the diagrams of this chapter, and check the results you have found.

# 7 Coordinates

The simplest sort of graph is that of a straight line. The equation of a straight line is of the form $y = mx + c$, where $m$ and $c$ are constant.

## Gradient and intercepts

- The constant $m$ is the **gradient** or slope, which is the $y$-change between a pair of points on the line, divided by the $x$-change.
- The constant $c$ is the value of $y$ when $x = 0$, i.e. it gives the point where the line crosses the $y$-axis. This is the **y-intercept**.

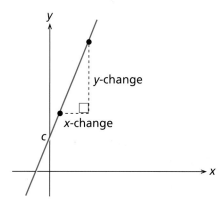

The **x-intercept** is the value of $x$ when $y = 0$. So put $y = 0$ into the equation, and solve to find the value of $x$.

For $y = mx + c$, put $y = 0$.

$$0 = mx + c$$
$$x = -\frac{c}{m}$$

For example, for the equation $y = 3x - 2$:

- the gradient is 3
- the $y$-intercept is $-2$
- the $x$-intercept is $\frac{2}{3}$.

## Exercise 7.1

Mainly revision

1. Find the gradients of the lines with the following equations.
   - **a** $y = 2x + 3$
   - **b** $y = -x + 7$
   - **c** $2y = x - 3$
   - **d** $3y = 2x + 4$
   - **e** $y + 3x = 4$
   - **f** $y - 3x = 5$
   - **g** $2y + 3x = 1$
   - **h** $4y - 3x = -3$
   - **i** $\frac{y}{3} + \frac{x}{4} = 5$

2. Find the $y$-intercepts of the equations in question 1.
3. Find the $x$-intercepts of the equations in question 1.

You can find the equation of a straight line if you know either of the following:

- the gradient and one point on it
- two points on it.

*Examples*   Find the equation of the straight line with gradient 3 which goes through (1, 7).

> So you know the gradient and one point on the line.

As $m = 3$, we know that the equation is of the form $y = 3x + c$. To find $c$, put $x = 1$ and $y = 7$.

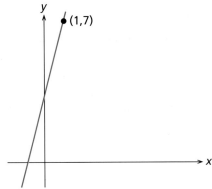

$$7 = 3 \times 1 + c$$
$$c = 4$$

The equation is $y = 3x + 4$.

**Note.** Another way of finding the equation is as follows. We know that the line goes through (1, 7) and has gradient 3. Hence any point $(x, y)$ on the line obeys

$$y - 7 = 3(x - 1)$$

Hence $y - 7 = 3x - 3$, giving $y = 3x + 4$.
This is the same answer as before.

---

> So you know two points on the line.

Find the equation of the line through (3, 1) and (−2, 5).

First find the gradient $m$.

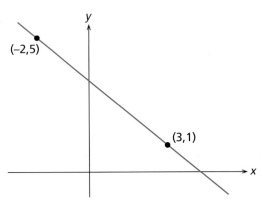

The $y$-change going from (3, 1) to (−2, 5) is $5 - 1$, i.e. 4.
The $x$-change going from (3, 1) to (−2, 5) is $-2 - 3$, i.e. $-5$.

So the gradient is $\dfrac{y\text{-change}}{x\text{-change}} = \dfrac{4}{-5} = -\tfrac{4}{5}$.

The equation is $y = -\tfrac{4}{5}x + c$. The line goes through (3, 1), so put $x = 3$ and $y = 1$.

$$1 = -\tfrac{4}{5} \times 3 + c$$
$$c = 1 + \tfrac{12}{5} = 3\tfrac{2}{5}$$

The equation is $y = -\tfrac{4}{5}x + 3\tfrac{2}{5}$.

Check: we used (3, 1) to find $c$. We can use the other point, $(-2, 5)$, to check the answer. Put $x = -2$ into the right-hand side of the equation.

$$-\tfrac{4}{5} \times (-2) + 3\tfrac{2}{5} = 1\tfrac{3}{5} + 3\tfrac{2}{5} = 5$$

This gives the correct value of $y$.

**Note.** It may look tidier to eliminate the fractions, by multiplying both sides of the equation by 5. The equation becomes

$$5y = -4x + 17$$

Find where the lines $y = 3x - 1$ and $y = 2x + 3$ meet.

The diagram shows the two lines.

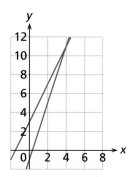

We want the point $(x, y)$ for which both equations are true, i.e. the point which satisfies both equations simultaneously. Equate the two expressions for $y$.

$$3x - 1 = 2x + 3$$
$$x = 4$$

If $x = 4$, then $y = 3 \times 4 - 1 = 11$ (using the first equation).
The lines meet at $(4, 11)$.

## Exercise 7.2

**1** Find the equation of each of the following lines.
 **a** Line with gradient 4 and through $(1, 4)$
 **b** Line with gradient $-3$ and through $(2, 1)$
 **c** Line with gradient $\tfrac{1}{3}$ and through $(6, -5)$
 **d** Line with gradient $-\tfrac{3}{4}$ and through $(2, 5)$
 **e** Line through $(1, 1)$ and $(4, 7)$
 **f** Line through $(5, 2)$ and $(2, 12)$
 **g** Line through $(-1, 3)$ and $(4, -2)$

**2** Find where each of these pairs of lines intersect.
 **a** $y = 3x - 5$ and $y = x + 3$
 **b** $y = 2x + 2$ and $y = -x + 11$
 **c** $y = \tfrac{1}{2}x + 3$ and $y = \tfrac{1}{6}x - 1$
 **d** $y = \tfrac{1}{4}x + 1$ and $y = -\tfrac{1}{3}x + 8$

# Parallel and perpendicular lines

The gradients of lines can often tell us geometric facts about them.

● *Two lines are parallel if they have the same gradient.*

So if $y = mx + c$ and $y = m'x + c'$ are parallel, then $m = m'$.

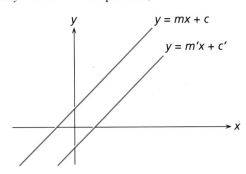

For example, the lines $y = 2x - 5$ and $y = 2x + 3$ are parallel.

**Proof**

The gradient is the slope of the line. If two lines have the same gradient, then they have the same slope, so $m = m'$.

The next result is less obvious.

● *If two lines are perpendicular, then the product of their gradients is $-1$.*

So if $y = mx + c$ and $y = m'x + c'$ are perpendicular, then $m \times m' = -1$.
For example, the lines $y = -2x + 5$ and $y = \frac{1}{2}x - 7$ are perpendicular.

**Proof**

The diagram shows two lines which are perpendicular to each other. For line 1, the gradient is $\dfrac{AB}{BX}$.

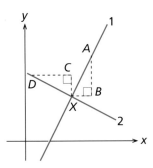

Line 2 can be obtained from line 1 by rotating it through 90°. Then the horizontal line $BX$ goes to the vertical line $CX$. The vertical line $AB$ goes to the horizontal line $CD$. Line 2 has a negative gradient, and hence

$$\text{gradient of line 2} = -\dfrac{CX}{CD} = -\dfrac{BX}{AB}$$

$$\text{gradient of line 1} \times \text{gradient of line 2} = \dfrac{AB}{BX} \times -\dfrac{BX}{AB} = -1.$$

*Example* Find the equations of the lines which are
   **a** parallel to $y = 3x - 7$ and through $(1, 4)$
   **b** perpendicular to $y = 4x - 3$ and through $(-2, 8)$.

   **a** If the line is parallel to $y = 3x - 7$, then its gradient $m$ is also 3. The equation is $y = 3x + c$. Put $x = 1$ and $y = 4$, giving $4 = 3 \times 1 + c$. Hence $c = 1$. The equation is $y = 3x + 1$.
   **b** Suppose the line is $y = mx + c$. Then, as the line is perpendicular to $y = 4x - 3$, which has gradient 4, we must have $m \times 4 = -1$, giving $m = -\frac{1}{4}$. The equation is $y = -\frac{1}{4}x + c$. Put $x = -2$ and $y = 8$.

   $$8 = -\tfrac{1}{4} \times (-2) + c$$
   $$c = 7\tfrac{1}{2}$$

   The equation is $y = -\frac{1}{4}x + 7\frac{1}{2}$.

**Note.** We can multiply this equation by 4, so that all the numbers are integers: $4y = -x + 30$.

# Exercise 7.3

**1** Which of the lines below are parallel or perpendicular to each other?
   **a** $y = 2x + 3$
   **b** $y = \frac{1}{2}x - 7$
   **c** $y = -2x + 1$
   **d** $y = -\frac{1}{2}x + 9$
   **e** $2y = 4x - 3$
   **f** $4y + 8x = 1$
   **g** $2y + x = 3$
   **h** $4y - 2x = 11$
**2** Find the equation of each of these lines.
   **a** Line parallel to $y = 2x - 1$ and through $(2, 5)$
   **b** Line parallel to $y = -\frac{1}{2}x$ and through $(4, 7)$
   **c** Line perpendicular to $y = 2x - 1$ and through $(4, 1)$
   **d** Line perpendicular to $y = \frac{1}{3}x + 2$ and through $(3, 8)$
**3** The lines $y = 2x + 7$ and $y = kx - 3$ are perpendicular. Find $k$.
**4** The lines $y = 3x + 1$ and $ky = x - 2$ are perpendicular. Find $k$.
**5** The lines $y = -kx + 1$ and $y = 4kx - 3$ are perpendicular. Find the possible values of $k$.
**6** The lines $y = kx + 3$ and $ky = 9x - 2$ are parallel. Find the possible values of $k$.
**7** Find the equation of the line perpendicular to the line $y = 2x + 3$ and through $(1, 7)$. Find where these two lines meet.
**8** Find the equation of the line perpendicular to the line $y = -\frac{3}{2}x + 1$ and through $(4, -3)$. Find where these two lines meet.

*Example* A quadrilateral has vertices at $A(1, 1)$, $B(5, 7)$, $C(8, 5)$ and $D(4, -1)$. Show that $ABCD$ is a rectangle.

The gradient of $AB$ is $\dfrac{7-1}{5-1} = \dfrac{6}{4} = \dfrac{3}{2}$. Similarly the gradients of $BC$, $CD$ and $DA$ are $-\tfrac{2}{3}, \tfrac{3}{2}$ and $-\tfrac{2}{3}$ respectively.

$\tfrac{3}{2} \times (-\tfrac{2}{3}) = -1$, hence $AB$ and $BC$ are perpendicular. Similarly the other pairs of adjacent sides are perpendicular.
$ABCD$ is a rectangle.
The diagram shows $ABCD$.

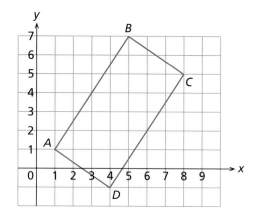

## Exercise 7.4

1. Find the gradients of the diagonals $AC$ and $BD$ in the example above. Show that the diagonals are not perpendicular (and hence that $ABCD$ is not a square).
2. A triangle has vertices at $A(4, 3)$, $B(6, 0)$, $C(12, 4)$. Show that the triangle is right-angled.
3. A quadrilateral has vertices at $A(0, 2)$, $B(4, 5)$, $C(9, 5)$ and $D(5, 2)$.
   a Show that opposite sides of $ABCD$ are parallel.
   b Show that the adjacent sides of $ABCD$ are not perpendicular.
   c Show that the diagonals of $ABCD$ are perpendicular.
   d What sort of quadrilateral is $ABCD$?
4. Three points are $A(2, 5)$, $B(5, 9)$ and $C(11, 17)$. Show that $AB$ is parallel to $BC$, and hence that $ABC$ is a straight line.
5. The points $A(1, 1)$, $B(2, 7)$ and $C(4, k)$ are in a straight line. Find $k$.
6. A quadrilateral has vertices at $A(1, 1)$, $B(4, 3)$, $C(7, 7)$ and $D(3, 4)$.
   a Are the opposite sides of $ABCD$ parallel?
   b Are the adjacent sides of $ABCD$ perpendicular?
   c Are the diagonals of $ABCD$ perpendicular?

# Distance between points

If you know the coordinates of two points $A$ and $B$, then you can use Pythagoras' theorem to find the distance between them. The diagram shows $A$ at $(4, 5)$ and $B$ at $(1, 1)$. The point $C$ is at $(4, 1)$.

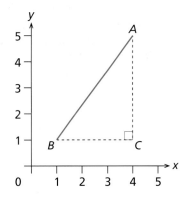

The horizontal distance between $A$ and $B$ is $BC$, which is 3 units. The vertical distance between $A$ and $B$ is $AC$, which is 4 units. Note that $\triangle ABC$ is right-angled at $C$. Hence

$$AB^2 = 3^2 + 4^2 = 25$$

The distance $AB$ is 5 units.

- In general, suppose that $A$ is at $(x_1, y_1)$ and $B$ is at $(x_2, y_2)$. Then the distance $AB$ is
$$\sqrt{(x_1 - x_2)^2 + (y_1 - y_2)^2}.$$

**Proof**

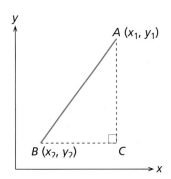

The horizontal distance between the points is the difference between the $x$-coordinates, which is $(x_1 - x_2)$. The vertical distance between the points is the difference between the $y$-coordinates, which is $(y_1 - y_2)$.

Note that $A$, $B$ and $C$ form a triangle which is right-angled at $C$. The hypotenuse of this triangle is $AB$. Pythagoras' theorem gives:

$$AB^2 = AC^2 + BC^2$$
$$AB^2 = (x_1 - x_2)^2 + (y_1 - y_2)^2$$

So $AB = \sqrt{(x_1 - x_2)^2 + (y_1 - y_2)^2}$, as required.

**Notes**

1 It does not matter whether we subtract the coordinates of $A$ from those of $B$ or vice versa. As we are squaring the differences, the result is positive, regardless of whether they are positive or negative. But be careful with the arithmetic of negative numbers, when some of the coordinates are themselves negative.
2 In both diagrams we showed $A$ above and to the right of $B$. The formula still holds if $A$ is below or to the left of $B$, as shown.

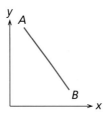 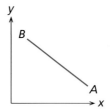 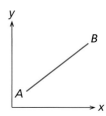

*Examples*   Find the distance between the points $(3, -6)$ and $(-5, -2)$.

Use the formula.
The difference between the $x$-coordinates is $(3 - (-5)) = (3 + 5) = 8$.
The difference between the $y$-coordinates is $(-6 - (-2)) = (-6 + 2) = -4$.

$$\text{The distance is } \sqrt{(8)^2 + (-4)^2} = \sqrt{64 + 16} = \sqrt{80}.$$

The distance is 8.94 units.

Let $A$, $B$ and $C$ be at $(1, 2)$, $(8, 7)$ and $(18, -7)$ respectively. Find the lengths of the sides of $\triangle ABC$. Show that the triangle is right-angled.

Use the formula for the lengths of the sides.

$$AB = \sqrt{7^2 + 5^2} = \sqrt{74}$$
$$BC = \sqrt{10^2 + 14^2} = \sqrt{296}$$
$$CA = \sqrt{17^2 + 9^2} = \sqrt{370}$$

Note that $74 + 296 = 370$. Hence the lengths of the sides obey Pythagoras' theorem.

$$AB^2 + BC^2 = CA^2$$

The triangle is right-angled.

**Note.** We could also show that $\triangle ABC$ is right-angled by the method of the previous section. The gradients of $AB$ and $BC$ are $\frac{5}{7}$ and $-\frac{14}{10}$ respectively. The product of these gradients is $-1$, which shows that $AB$ and $BC$ are perpendicular.

# Exercise 7.5

1 Find the distance between each of the following pairs of points.
   **a** $(1, 2)$ and $(5, 7)$
   **b** $(2, 8)$ and $(3, 6)$
   **c** $(-2, 5)$ and $(3, 10)$
   **d** $(4, 1)$ and $(-4, 8)$
   **e** $(-3, 7)$ and $(-5, 1)$
   **f** $(2, -8)$ and $(9, -4)$
   **g** $(-3, -9)$ and $(7, -2)$
   **h** $(7, -11)$ and $(-6, 4)$

2 The three vertices of a triangle are at (2, 3), (5, −2) and (7, 6). Find the lengths of the sides of the triangle, and show that it is right-angled.

3 The three vertices of a triangle are at (7, 1), (−3, 5) and (3, 20). Find the lengths of the sides of the triangle, and show that it is right-angled.

4 Show that the triangle with vertices at (1, 1), (6, 6) and (2, 8) is isosceles. Is it equilateral?

5 Show that the triangle with vertices at (19, 18), (4, −2) and (11, 22) is isosceles.

6 Three points are $A(2, 3)$, $B(6, 6)$ and $C(14, 12)$. Find the lengths of $AB$, $BC$ and $AC$. Hence show that $A$, $B$ and $C$ are in a straight line.

7 By considering lengths, show that (1, 1), (7, 10) and (11, 16) are in a straight line.

8 The vertices of a quadrilateral are at $A(3, 4)$, $B(6, 8)$, $C(9, 7)$ and $D(8, 4)$. Find the lengths of $AB$, $AD$, $CB$ and $CD$. Show that the quadrilateral is a kite.

9 The vertices of a quadrilateral are at (2, 3), (5, 9), (−1, 12) and (−4, 6). Show that the quadrilateral is a rhombus. Is it a square?

10 A circle has centre (1, 1) and radius 5 units. Which of the following points lie on the circle?
   a (4, 5)
   b (−3, 4)
   c (3, 4)
   d (1, 6)
   e (6, −4)

11 $A$, $B$, $C$ and $D$ are at (1, −2), (5, 6), (3, 2) and (−1, 4) respectively.
   a Show that $C$ is the midpoint of $AB$.
   b Show that $D$ lies on the circle with $AB$ as diameter.

12 $A$ is at (0, 1) and $B$ is at (4, 5). $P$ is a point for which $PA = PB$. If $P$ is at $(x, y)$, show that $x + y = 5$.

13 The point $A(x, 2)$ is at a distance of 5 units from $B(1, 5)$. Find the two possible values of $x$.

14 The point $(x, x)$ is a distance of 13 units from (8, 1). Find the two possible values of $x$.

# Simultaneous linear and quadratic equations

You have already solved linear simultaneous equations, i.e. a pair of equations such as

$$3x + 2y = 15$$
$$4x - 3y = 21$$

The method involved is **elimination**.

If one of the equations is quadratic, i.e. involves $x^2$, then you may be able to solve them by **substitution**, i.e. by using one of the equations to write one of the variables in terms of the other and substituting it in the other equation. Use the linear equation to write one variable in terms of the other, and then substitute in the quadratic equation.

**Remember:**
You eliminate one variable, so that you are left with an equation in just the other variable.

## 126 COORDINATES

*Example* Solve the simultaneous equations
$$3x + 2y = 22$$
$$y = x^2 + 3x - 2$$

Use the first equation to write $y$ in terms of $x$.
$$2y = 22 - 3x$$
$$y = 11 - 1.5x$$

Now substitute into the second equation.
$$11 - 1.5x = x^2 + 3x - 2$$
$$x^2 + 4.5x - 13 = 0$$
$$2x^2 + 9x - 26 = 0 \quad \text{(doubling to get rid of the fraction)}$$

This factorises to
$$(x - 2)(2x + 13) = 0$$

Hence $x = 2$ or $x = -6.5$.
  If $x = 2$, then $y = 11 - 1.5 \times 2 = 8$.
  If $x = -6.5$, then $y = 11 - 1.5 \times (-6.5) = 20.75$.
Either $x = 2$ and $y = 8$, or $x = -6.5$ and $y = 20.75$.

The diagram shows the graphs of $3x + 2y = 22$ and $y = x^2 + 3x - 2$. Note that they cross at two points, $(2, 8)$ and $(-6.5, 20.75)$. But by graphical methods it is very difficult to achieve this degree of accuracy.

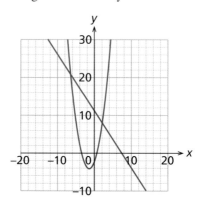

**Note.** We solved the quadratic by factorisation. If the quadratic had not factorised, then we would have had to use the formula.

# Exercise 7.6

In questions 1–12, solve the simultaneous equations. Where relevant, give your answers correct to three significant figures.

1. $y = x^2$ and $y = 3x - 2$
2. $y = x^2$ and $y = 2 - x$
3. $y = x^2 + x$ and $y = 2x + 1$
4. $y = x^2 + 3x - 2$ and $y = 2x + 2$

5. $y = x^2 - 2x - 5$ and $y = x - 2$
6. $y = 3 - x^2$ and $y = 2x - 1$
7. $y = 2 - 2x - x^2$ and $y = x + 1$
8. $y = 2x^2 + x + 2$ and $y = 3x + 3$
9. $y = 3x^2 - 2x - 4$ and $y = 3 - 2x$
10. $y = 6 + x - 2x^2$ and $y = 2x - 3$
11. $y = -1 + 3x - 3x^2$ and $y = x - 3$
12. $y = -2 - 5x - 2x^2$ and $y = 3x - 4$
13. Solve the simultaneous equations $y = x^2$ and $y = 2x - 1$. What is the relationship between the curve and the straight line?

# Circles

The diagram shows a circle with centre $(0, 0)$ and radius 2 units. Suppose a typical point $P$ on the circle is $(x, y)$. Then the horizontal distance from $P$ to $(0, 0)$ is $x$, and the vertical distance is $y$. Since the length of $OP$ is always 2, we have by Pythagoras' theorem

$$x^2 + y^2 = 2^2 = 4$$

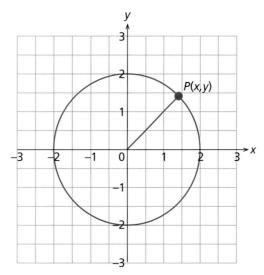

This is the equation of the circle. In general, if a circle has centre $(0, 0)$ and radius $k$, then its equation is

$$x^2 + y^2 = k^2$$

Two non-parallel lines meet in one point. But a line and a circle may not meet at all or meet at one or two points. This diagram shows the three cases.

 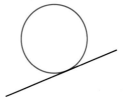 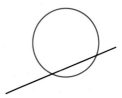

# 128 COORDINATES

*Example*  Draw the circle with equation $x^2 + y^2 = 9$. Draw the line with equation $y = 2x - 1$. Find the coordinates of the points where the line crosses the circle.

The circle has a centre (0, 0) and radius 3. It is shown in red.

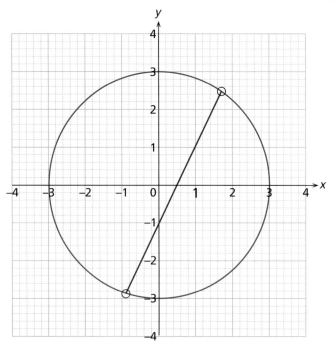

For the line $y = 2x - 1$, put $x = 0$ and then $y = 0$, to find that the line meets the axes at $(0, -1)$ and $(\frac{1}{2}, 0)$. Join these points up to draw the line (blue). The crossing points are shown ringed. Read off the coordinates.
The line meets the circle at $(-0.9, -2.9)$ and $(1.7, 2.5)$.

## Exercise 7.7

**You will need:**
- graph paper
- compasses
- ruler

1  Draw the circle with equation $x^2 + y^2 = 1$. Draw the line with equation $y = 3x - 2$. Write down the coordinates of the points where the line meets the circle.
2  Draw the circle with equation $x^2 + y^2 = 4$. Draw the line with equation $y = 3 - 2x$. Write down the coordinates of the points where the line meets the circle.
3  Draw the circle with equation $x^2 + y^2 = 2$. Draw the line with equation $y = x - 1$. Write down the coordinates of the points where the line meets the curve.
4  On your graph of $x^2 + y^2 = 2$, draw the line $y = 2x - 1$. Write down the coordinates of the points where the line meets the curve.

In this section we have found the intersection of a line and a circle by drawing. It is quicker and more accurate to do it by calculation, as we shall now see.

*Example*  Solve the equations

$$y = 2x - 1 \qquad x^2 + y^2 = 9$$

**Note.** These are the equations of the line and the circle in the example above. Here we are finding the intersection points by calculation rather than by drawing.

From the first equation, $y = 2x - 1$. Substitute this in the second equation.

$$x^2 + (2x - 1)^2 = 9$$
$$x^2 + 4x^2 - 4x + 1 = 9$$
$$5x^2 - 4x - 8 = 0$$

$$x = \frac{-(-4) \pm \sqrt{4^2 - 4 \times 5 \times (-8)}}{2 \times 5}$$

$$= \frac{4 \pm \sqrt{176}}{10}$$

$x = -0.927$ or $x = 1.727$.

Now find the corresponding values of $y$.
  For $x = -0.927$, $y = 2 \times (-0.927) - 1 = -2.853$.
  For $x = 1.727$, $y = 2 \times 1.727 - 1 = 2.453$.
The solutions are $x = -0.927$ and $y = -2.853$, or $x = 1.727$ and $y = 2.453$.

> Compare these results with those obtained by drawing on page 128. The results obtained by calculation are more accurate.

# Exercise 7.8

Solve the following simultaneous equations.

1  $y = 2x - 1$ and $x^2 + y^2 = 3$
2  $y + x = 6$ and $x^2 + y^2 = 24$
3  $2y + 3x = 12$ and $x^2 + y^2 = 13$
4  $y = 3x + 2$ and $x^2 + y^2 = 28$
5  $2y + x = 7$ and $x^2 + y^2 = 13$
6  $3y = 2x - 1$ and $x^2 + y^2 = 16$
7  $y = 2x + 1$ and $x^2 + y^2 = 7$
8  $y + x = 3$ and $y^2 - x^2 = 5$
9  $2y + 3x = 5$ and $x^2 + 2y^2 = 11$
10  $y = x + 3$ and $x^2 + x + y^2 = 7$
11  $y = 5 - 2x$ and $2x^2 + 3y + y^2 = 12$
12  $y = 4x - 1$ and $3x^2 + xy + y^2 = 27$

# SUMMARY

- You can find the equation of a straight line if you know its **gradient** and one point on it, or if you know two points on it.
- Suppose two lines are $y = mx + c$ and $y = m'x + c'$. The lines are parallel if $m = m'$, and they are perpendicular if $mm' = -1$.
- If $A$ is at $(x_1, y_1)$ and $B$ is at $(x_2, y_2)$, then the distance $AB$ is $\sqrt{(x_1 - x_2)^2 + (y_1 - y_2)^2}$.
- The intersection of a straight line and a quadratic curve can be found by calculation (**substitution**) as well as by drawing.
- The circle with centre $(0, 0)$ and radius $k$ has equation $x^2 + y^2 = k^2$.
- A line and a curve with equation involving $x^2$ or $y^2$ may not meet at all, or meet at one or two points. These points can be found either by drawing or by solving equations.

# 130 COORDINATES

## Exercise 7A

1. For the line with equation $y = 4x - 3$, find the gradient, the $y$-intercept and the $x$-intercept.
2. Find the equation of the straight line through $(3, 1)$ and $(4, -5)$.
3. If $y = 3x + 7$ and $y = kx - 2$ are perpendicular, find $k$.
4. Find the equation of the straight line through $(4, -3)$ which is perpendicular to $y = 2x - 3$.
5. A quadrilateral has vertices at $A(1, 2)$, $B(3, 5)$, $C(7, 6)$ and $D(9, 4)$. Show that $ABCD$ is a trapezium but not a parallelogram.
6. A triangle has vertices at $(4, 2)$, $(5, 7)$ and $(2, 5)$. Show that the triangle is right-angled.
7. Find $h$, given that the points $(1, 2)$, $(3, 5)$ and $(7, h)$ are in a straight line.
8. Find the distance between the points $(4, 8)$ and $(-3, 7)$.
9. Find the solution of the simultaneous equations $2y - 3x = 4$ and $y = x^2 + x - 4$.
10. Draw the circle with equation $x^2 + y^2 = 8$. On your diagram draw the line with equation $y = 2x - 3$. Write down the coordinates of the points where the line meets the circle.

**You will need:**
- graph paper
- compasses
- ruler

## Exercise 7B

1. For the line with equation $2y + 3x = 12$, find the gradient, the $y$-intercept and the $x$-intercept.
2. Find the equation of the straight line which has gradient $\frac{3}{4}$ and which goes through $(8, -7)$.
3. If $2hy = x + 3$ and $y = 18hx - 3$ are parallel, find the possible values of $h$.
4. Find the equation of the straight line through $(1, 7)$ which is perpendicular to $3y + 2x = 5$.
5. Find where the two lines of question 4 meet.
6. A quadrilateral has vertices at $A(1, 1)$, $B(4, -3)$, $C(12, 3)$ and $D(9, 7)$. Show that $ABCD$ is a rectangle but not a square.
7. Find $h$, given that the points $(1, 3)$, $(h, h)$ and $(7, 5)$ are in a straight line.
8. A quadrilateral has vertices at $(0, 0)$, $(3, 4)$, $(9, 9)$ and $(4, 3)$. Show that the quadrilateral is a kite.
9. Calculate the points where the line $y = 3x - 1$ meets the curve $y = 2 + x - x^2$.
10. Solve the simultaneous equations $y + 3x = 2$ and $2x^2 + y^2 = 7$.

## Exercise 7C

There are many centres associated with triangles. Now you can find their positions by calculation instead of by drawing.

Take a triangle with vertices at $A(2, 8)$, $B(8, 12)$ and $C(8, 2)$.

1. The **orthocentre** is found by drawing perpendiculars from each vertex to the opposite sides. Find the equations of the perpendiculars from $B$ to $AC$ and from $C$ to $AB$. Find the intersection of these perpendiculars. This is the orthocentre.

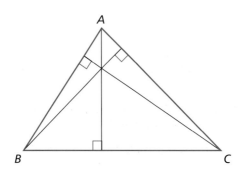

2 The **circumcentre** is the centre of the circle which goes through the vertices of the triangle. It is where the perpendicular bisectors of the sides meet. Find the midpoints of $AC$ and $AB$ (by averaging their coordinates) and hence find the equations of the perpendicular bisectors. Find the intersection of these bisectors.

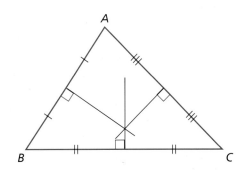

3 The **centroid** is where the lines joining the vertices to the midpoints of the opposite sides meet. Find the equations of the lines joining $B$ to the midpoint of $AC$, and $C$ to the midpoint of $AB$. Find the intersection of these lines.

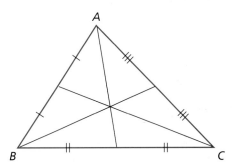

4 Show that these three centres lie on a straight line.

> This straight line is known as the Euler line. You read about Leonhard Euler in chapter 5.

# Exercise 7D

**You will need:**
- graph paper
- compasses
- ruler

This exercise mixes the algebra of this chapter with geometrical constructions. You find the equation of a tangent from a point to a line.

1 Draw the circle with equation $x^2 + y^2 = 1$. Take any point $P$ outside the circle. Join $P$ to the origin $O$. Bisect this line, and draw the circle with $OP$ as diameter. Let this circle cut the original circle at $A$ and $B$. Join $PA$.
2 Show that $PA$ is a tangent to the first circle. Refer to chapter 5.
3 Write down the coordinates of $A$, and find the gradients of $OA$ and $PA$. Are $OA$ and $PA$ perpendicular?

# Exercise 7E

**You will need:**
- graph-plotting program or graphics calculator capable of plotting graphs such as $x^2 + y^2 = k^2$

The equation of a circle with centre $O$ and radius $k$ is $x^2 + y^2 = k^2$. In this exercise you see what curves are obtained if the equation is altered.

1 What if the coefficient of $x^2$ or $y^2$ is not 1? Plot the graphs of $2x^2 + y^2 = 1$, $3x^2 + 2y^2 = 6$ and similar expressions. What do they look like? What are the intercepts with the axes?

2 What if there is an $xy$ term in the equation? Plot the graph of $x^2 + xy + y^2 = 1$. What does it look like? How does it compare with your answers to question 1?

3 Plot the graph of $x^2 + kxy + y^2 = 1$, for various values of $k$, starting with $k = 0$ and increasing $k$ up to 4. What do the curves look like?

# 8: Transformations of graphs

You have already drawn many graphs, some of which had similar shapes. Look at these two graphs, of $y = x^2$ and $y = x^2 + 2$.

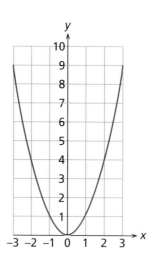 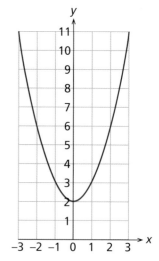

The two graphs have exactly the same shape. The second graph can be obtained from the first by raising all the points by 2 units. The word to describe this transformation is **translation**. The second graph is obtained from the first by a translation of 2 units upwards, i.e. by the vector $\binom{0}{2}$.

In this chapter we show how graphs can be obtained from each other by transformations.

## Translating graphs

The graph of $y = x^2 + 2$ above is obtained from the graph of $y = x^2$ by a translation of 2 upwards. Similarly the graph of $y = x^2 - 3$ is obtained from the graph of $y = x^2$ by a translation of 3 downwards, i.e. by the vector $\binom{0}{-3}$.

For any constant $a$, the graph of $y = x^2 + a$ is obtained from the graph of $y = x^2$ by a translation of $a$ units upwards. If $a$ is negative, the graph is moved downwards.

This result is true for any graph, whatever its function. In general, if we add $a$ to all the $y$-coordinates of the points on a graph, the whole graph is translated $a$ upwards. (If $a$ is negative, then the graph is translated downwards.)

We write the equation of a general graph as $y = f(x)$. The letter f stands for any function which could be applied to $x$. So for example:

for $y = x^2$, $f(x) = x^2$
for $y = \cos x$, $f(x) = \cos x$

So suppose the general graph has equation $y = f(x)$. Then adding $a$ to all the $y$-coordinates converts the equation to $y = f(x) + a$. The whole graph is shifted by $a$ upwards. The diagram shows the two graphs.

## Translating graphs

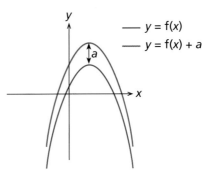

In particular, important features of the graph, such as any maximum or minimum, or the $y$-intercept, are moved upwards by $a$.

For example, look at these graphs of $y = x^2 - 2x$ and $y = x^2 - 2x + 2$. The first has a minimum at $(1, -1)$, and the second at $(1, 1)$. So the minimum has been shifted up by 2.

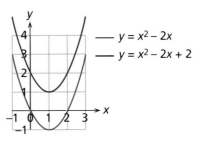

*Example*

> If we do not know the function $f(x)$, we do not know the equation of the graph. A graphics calculator is no use if you are given a general $f(x)$!

The graph of $y = f(x)$ is shown. Sketch the graph of $y = f(x) + a$, where $a > 0$. Give the coordinates of the new maximum point, and of the $y$-intercept.

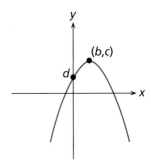

Every point on the graph is raised by $a$.

The old maximum was at $(b, c)$. The new maximum has its $y$-value increased by $a$, but its $x$-value is unchanged.

The new maximum is at $(b, c + a)$.

The old intercept is at $(0, d)$. Move this up the $y$-axis $a$ units.
The new $y$ intercept is at $(0, d + a)$.

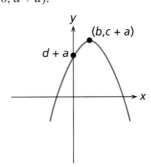

## Exercise 8.1

**You will need:**
- squared paper

1. Use the graph of $y = x^2$ to sketch the following graphs. In each case write down the coordinates of the minimum.
   a  $y = x^2 + 2$
   b  $y = x^2 - 1$

2. The diagram shows the graph of $y = f(x)$.

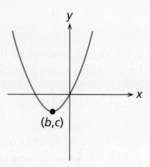

   a  Sketch the graph of $y = f(x) + a$, where $a$ is positive.
   b  Write down the coordinates of the new minimum.
   c  Write down the coordinates of the new $y$-intercept.

3. The diagram shows the graph of $y = f(x)$.

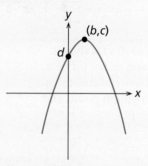

   a  Sketch the graph of $y = f(x) + a$, where $a$ is negative.
   b  Write down the coordinates of the new maximum.
   c  Write down the coordinates of the new $y$-intercept.

4. The graph of $y = 2x$ is shown. The graph crosses the $y$-axis at $(0, 0)$. Sketch the graphs of the following. In each case write down the coordinates of the $y$-intercept.

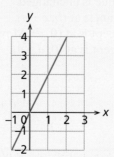

   a  $y = 2x + k$
   b  $y = 2x - k$

5 The graph of $y = \sin x$ is shown. It reaches a peak of 1 at 90°, and it crosses the y-axis at (0, 0). Sketch the graphs of the following. In each case write down the coordinates of the new peak and of the y-intercept.

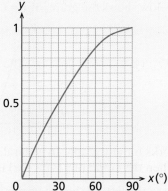

a $y = \sin x + 2$
b $y = \sin x - 1$.

6 The diagram shows the graph of $y = 2x - x^2$. It has a highest point at (1, 1) and it crosses the y-axis at (0, 0). Sketch the graph of the following, writing down the coordinates of the highest point and of the y-intercept.

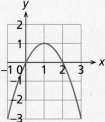

a $y = 2x - x^2 + 2$
b $y = 2x - x^2 - 3$

**Remember:**
*The right–left movement (change in x-coordinate) is on the top of the vector, and the up–down movement (change in y-coordinate) is on the bottom.*

Look at the following two graphs, which are of $y = x^2$ and $y = (x+2)^2$. They have the same shape, but the second graph is obtained by moving the first graph 2 units to the left. The second graph is obtained from the first by a translation of 2 to the left, i.e. by the vector $\begin{pmatrix} -2 \\ 0 \end{pmatrix}$.

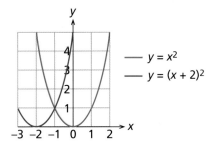

In particular, look at the lowest point of the graph. The lowest point of $y = x^2$ is at (0, 0). The lowest point of $y = (x+2)^2$ is where the expression inside the brackets is zero, i.e. where $x + 2 = 0$. So the lowest point of $y = (x+2)^2$ is at $(-2, 0)$.

## 136 TRANSFORMATIONS OF GRAPHS

For any $a$, the graph of $y = (x+a)^2$ can be obtained by shifting the graph of $y = x^2$ by $a$ units to the left. The lowest point of the graph is where $x + a = 0$, i.e. at $(-a, 0)$. If $a$ is negative then the graph is shifted $a$ units to the right. The diagram below shows the graph of $y = (x-3)^2$.

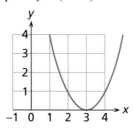

This works for any graph. In general, suppose we are given the graph of $y = f(x)$. Then the graph of $y = f(x+a)$ is obtained by translating the original graph by $a$ to the left, i.e. by the vector $\begin{pmatrix} -a \\ 0 \end{pmatrix}$. If $a$ is negative then the graph is translated to the right.

**Note.** It may seem peculiar that the result of *adding a* to $x$ has the effect of moving the graph to the *left*, and not to the right. When we add $a$ to $x$, points on the graph occur $a$ units *sooner* than on the original graph. That is why the graph moves to the left – because events happen $a$ units earlier.

*Examples*   Use the graph of $y = x^2$ to sketch the graph of $y = (x+3)^2$. Write down the coordinates of the lowest point of $y = (x+3)^2$.

The graph is shifted to the left by 3 units. The diagram shows $y = x^2$ and $y = (x+3)^2$.

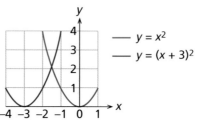

The lowest point of $y = x^2$ is at $(0, 0)$. Move this 3 units to the left. The lowest point of $y = (x+3)^2$ is at $(-3, 0)$.

The graph of $y = f(x)$ is shown. Sketch the graph of $y = f(x+a)$, where $a$ is positive. Give the coordinates of the maximum and the point where the graph crosses the $x$-axis.

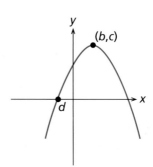

Move the graph *a* units to the left. The result is shown.

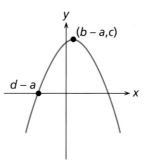

The *y*-coordinate of the maximum is unchanged, but the *x*-coordinate is reduced by *a*.
The maximum is at $(b - a, c)$.
The point where the graph crosses the *x*-axis is moved to the left by *a*.
The graph crosses the *x*-axis at $(c - a, 0)$.

## Exercise 8.2

**You will need:**
- squared paper

**1** Sketch the graphs of the following. In each case write down the coordinates of the lowest point.
   **a** $y = (x + 1)^2$
   **b** $y = (x - 4)^2$

**2** The graph of $y = f(x)$ is shown. Sketch the graph of $y = f(x + a)$, where *a* is positive.

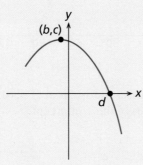

Write down the coordinates of the new maximum point, and of the *x*-intercept.

**3** The graph of $y = f(x)$ is shown. Sketch the graph of $y = f(x + a)$, where *a* is negative.

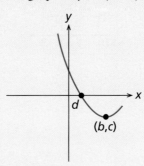

Write down the coordinates of the new minimum point, and of the *x*-intercept.

**4** The graph of $y = 4x - x^2$ is shown. It has a maximum at $(2, 4)$ and it crosses the $x$-axis at $(0, 0)$. Sketch the graphs of the following. In each case write down the coordinates of the maximum and where the graph crosses the $x$-axis.

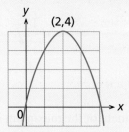

**a** $y = 4(x - 1) - (x - 1)^2$
**b** $y = 4(x + 1) - (x + 1)^2$

**5** The graph of $y = \cos x$ is shown. It reaches a peak of 1 at $0°$, and it crosses the $x$-axis at $(90°, 0)$. Sketch the graphs of the following. In each case write down the coordinates of the new peak and of the $x$-intercept.

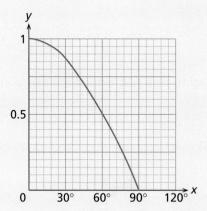

**a** $y = \cos(x + 30°)$
**b** $y = \cos(x - 60°)$

**6** The graph of $y = 3 - 2x$ is shown. It crosses the $x$-axis at $(1.5, 0)$. Sketch the graph of each of the following, writing down the coordinates of the $x$-intercept.

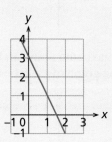

**a** $y = 3 - 2(x + k)$
**b** $y = 3 - 2(x - k)$

## Combined translations

We can do translations vertically and horizontally. For example, suppose we have the graph of $y = (x + a)^2 + b$. This is obtained from the graph of $y = x^2$ by a translation of $a$ to the left and $b$ upwards, i.e. by a translation by the vector $\begin{pmatrix} -a \\ b \end{pmatrix}$. The expression $y = (x + a)^2 + b$ is obtained when we complete the square of a quadratic expression.

*Example*  Complete the square of the expression $x^2 - 4x + 1$. Hence sketch the graph of the equation $y = x^2 - 4x + 1$.

> Use the technique of chapter 3.

$$x^2 - 4x + 1 = (x - 2)^2 - 3$$

So the equation becomes $y = (x - 2)^2 - 3$. The graph is obtained from that of $y = x^2$ by a translation of 2 to the right and 3 down, i.e. by the vector $\begin{pmatrix} 2 \\ -3 \end{pmatrix}$. The result is shown in the diagram.

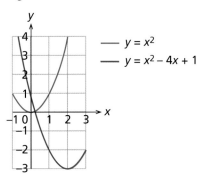

## Exercise 8.3

You will need:
- squared paper

1  Sketch the graphs of the following. In each case write down the coordinates of the maximum or minimum points.
   a  $y = (x + 2)^2 - 1$
   b  $y = (x - 3)^2 + 2$
   c  $y = 4 - (x - 2)^2$
   d  $y = -1 - (x + 3)^2$

2  By completing the square of the following expressions, sketch their graphs.
   a  $y = x^2 + 2x - 1$
   b  $y = x^2 - 4x - 1$
   c  $y = -x^2 + 2x + 3$
   d  $y = -x^2 - 6x - 1$

# Stretching graphs

Look at these graphs, of $y = x^2$ and $y = 2x^2$. They are similar in shape, but the $y$-values of the second graph have been increased by a factor of 2. The result is to 'stretch' the graph by a factor of 2 in the direction of the $y$-axis.

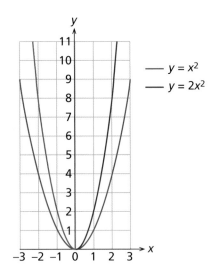

In general, given the graph of $y = f(x)$, the graph of $y = af(x)$, where $a > 1$, is obtained by 'stretching' the original graph by a factor of $a$, in the direction of the $y$-axis. Any point not on the $x$-axis is moved further from the $x$-axis by a factor of $a$.

If $0 < a < 1$, then the graph is 'shrunk' or 'compressed' in the direction of the $y$-axis.

*Example* The graph of $y = f(x)$ is shown. Sketch the graph of $y = af(x)$, where $a > 1$. Give the coordinates of the maximum and of the $y$-intercept.

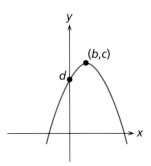

All the $y$-values move further from the $x$-axis, by a factor of $a$. The result is shown.

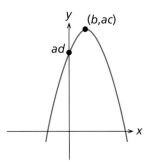

The maximum is now at $(b, ac)$.
The $y$-intercept is now at $(0, ad)$.

## Stretching graphs 141

## Exercise 8.4

**You will need:**
- squared paper

**1** Sketch the graphs of the following.
  **a** $y = 3x^2$
  **b** $y = \tfrac{1}{2}x^2$

**2** The graph of $y = f(x)$ is shown. Sketch the graph of $y = af(x)$, where $a > 1$. Write down the coordinates of the maximum point and of the $y$-intercept.

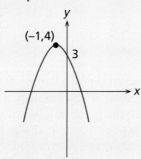

**3** The graph of $y = f(x)$ is shown. Sketch the graph of $y = af(x)$, where $0 < a < 1$. Write down the coordinates of the minimum point and of the $y$-intercept.

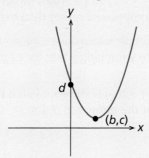

**4** Refer to the graph of $y = 2x - x^2$ on page 135. Sketch the graph of $y = 2(2x - x^2)$, writing down the coordinates of the maximum and of the $y$-intercept.

**5** Refer to the graph of $y = \cos x$ on page 138. Sketch the graph of $y = 2 \cos x$, writing down the coordinates of the maximum and of the $y$-intercept.

**6** Refer to the graph of $y = 3 - 2x$ on page 138. Sketch the graph of $y = 2(3 - 2x)$.

Look at these graphs of $y = x^2$ and $y = (2x)^2$. The second graph is 'narrower' than the first, by a factor of 2. The effect of changing $x$ to $2x$ makes things happen twice as quickly, so the shape of the graph is 'compressed' in the direction of the $x$-axis. If we multiply $x$ by $a$, where $a > 1$, the function changes to $y = (ax)^2$, which has a graph thinner by a factor of $a$.

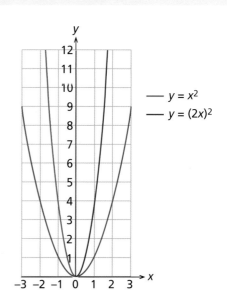

# 142 TRANSFORMATIONS OF GRAPHS

If $0<a<1$ the shape of the graph is made fatter. The graph of $y=(\frac{1}{2}x)^2$ is shown.

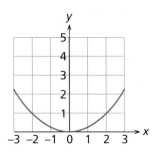

In general, given the graph of $y=f(x)$, the graph of $y=f(ax)$, where $a>1$, is found by 'compressing' the graph by a factor of $a$ in the direction of the $x$-axis. If $0<a<1$, then the graph will be stretched in the direction of the $x$-axis. In the graph of $y=f(\frac{1}{3}x)$, for example, things happen three times as slowly as in $y=f(x)$, hence the graph is stretched by a factor of 3 in the direction of the $x$-axis.

It may seem odd that multiplying $x$ by $a$ compresses the graph. When you change from $y=f(x)$ to $y=f(ax)$, things are happening $a$ times as quickly. Hence the graph is compressed rather than expanded.

**Note.** The example given above, of $y=(2x)^2$, can also be written as $y=4x^2$. This corresponds to stretching $y=x^2$ by a factor of 4 parallel to the $y$-axis.

*Example*  The graph of $y=f(x)$ is shown. Sketch the graph of $y=f(2x)$. Give the coordinates of the minimum point.

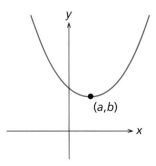

Compress the graph by a factor of 2 parallel to the $x$-axis. The result is shown. The minimum point moves closer to the $y$-axis, by a factor of 2.
The coordinates of the minimum point are $(\frac{1}{2}a, b)$.

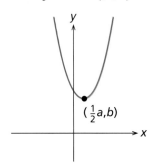

## Exercise 8.5

**You will need:**
- squared paper

1. Sketch the graph of each of the following.
   a. $y = (3x)^2$
   b. $y = (\frac{1}{3}x)^2$

2. The graph of $y = f(x)$ is shown. Sketch the graph of $y = f(ax)$, where $a > 1$. Write down the coordinates of the minimum point and of the $x$-intercepts.

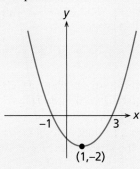

3. The graph of $y = f(x)$ is shown. Sketch the graph of $y = f(ax)$, where $0 < a < 1$. Write down the coordinates of the maximum point and of the $x$-intercepts.

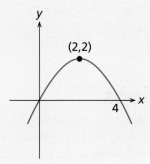

4. Refer to the graph of $y = 4x - x^2$ on page 138. Sketch the graph of $y = 4(2x) - (2x)^2$, writing down the coordinates of the maximum point and of the $x$-intercepts.

5. Refer to the graph of $y = \cos x$ on page 138. Sketch the graph of $y = \cos(\frac{1}{3}x)$, writing down the coordinates of the maximum and of the $x$-intercept.

6. The graph of $y = 3x - 1$ is shown. It crosses the $x$-axis at $(\frac{1}{3}, 0)$. Sketch the graphs of the following, writing down the coordinates of the $x$-intercept in each case.

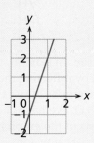

   a. $y = 3(ax) - 1$  $(a > 1)$
   b. $y = 3(ax) - 1$  $(0 < a < 1)$

## 144 TRANSFORMATIONS OF GRAPHS

**7** The graph of $y = x^2 - 4x - 5$ is shown. It has a minimum at $(2, -9)$ and it crosses the $x$-axis at $(-1, 0)$ and $(5, 0)$. Sketch the graphs of the following, writing down the coordinates of the minimum and of the $x$-intercepts.

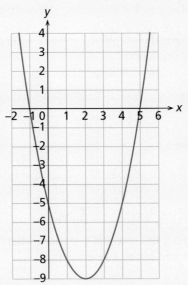

**a** $y = (3x)^2 - 4(3x) - 5$
**b** $y = (\frac{1}{2}x)^2 - 4(\frac{1}{2}x) - 5$

# Reflections

The diagram shows the graphs of $y = x^2 - x$ and $y = -x^2 + x$. The second graph is obtained by multiplying all the $y$-values of the first graph by $-1$. This result is the same as a **reflection** of the original graph in the $x$-axis.

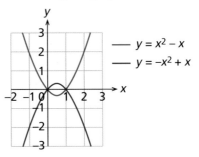

This is always true. In general, given the graph of $y = f(x)$, the graph of $y = -f(x)$ is found by reflecting the original graph in the $x$-axis.

Now look at the graphs of $y = x^2 - x$ and $y = (-x)^2 - (-x)$. We have changed $x$ to $-x$, so the parts of the graph for positive values of $x$ are now for negative values and vice versa. So the effect of changing $x$ to $-x$ is to swap around the positive and negative $x$-values, which is the same as reflecting the graph in the $y$-axis.

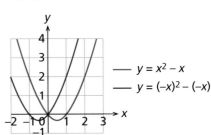

In general, given the graph of $y = f(x)$, the graph of $y = f(-x)$ is obtained by reflecting the original graph in the $y$-axis.

*Example*  The graph of $y = f(x)$ is shown. Sketch the graph of $y = -f(x)$ and of $y = f(-x)$. In each case state the coordinates of the points where the graph crosses the axes.

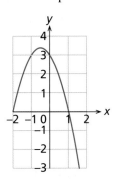

Reflect the graph in the $x$-axis. The result is shown.

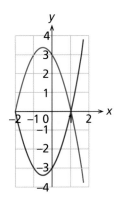

The $x$-intercepts are unchanged.
  The original $y$-intercept is $+3$. Hence the new $y$-intercept is $-3$.
The graph crosses the $y$-axis at $(0, -3)$.
  Reflect the original graph in the $y$-axis. The result is shown.

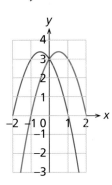

The $y$-intercept is unchanged.
  The $x$-intercepts were at $-2$ and $1$. Multiply these by $-1$.
The graph crosses the $x$-axis at $(-1, 0)$ and $(2, 0)$.

# 146 TRANSFORMATIONS OF GRAPHS

## Exercise 8.6

**You will need:**
- squared paper

1 Sketch the graph of $y = -x^2$.
2 Refer to the graph of $y = 2x - x^2$ on page 135. Sketch the graph of $y = -2x + x^2$. Write down the coordinates of the maximum or minimum.
3 The graph of $y = f(x)$ is shown. Sketch the graphs of the following. In each case write down the coordinates of the maximum or minimum and of the $x$- and $y$-intercepts.

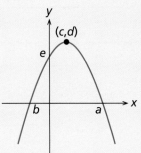

  **a** $y = -f(x)$
  **b** $y = f(-x)$

4 The graph of $y = f(x)$ is shown. Sketch the graphs of the following. In each case write down the coordinates of the maximum or minimum and of the $x$- and $y$-intercepts.

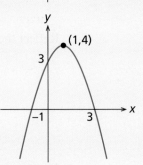

  **a** $y = -f(x)$
  **b** $y = f(-x)$

5 Refer to the graph of $y = \sin x$ on page 135. Sketch the graph of $y = -\sin x$.
6 Refer to the graph of $y = 4x - x^2$ on page 138. Sketch the graph of the following, in each case writing down the coordinates of the maximum or minimum.
  **a** $y = -4x + x^2$
  **b** $y = 4(-x) - (-x)^2$

## Combined transformations

Translations, stretches and reflections can be combined with each other. Be careful to do the transformations in the correct order.

*Example*    The diagram shows the graph of $y = f(x)$. Sketch the graphs of
  **a** $y = f(x + 2)$
  **b** $y = -f(x + 2)$

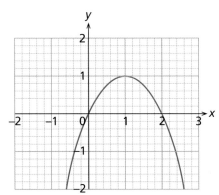

## Combined transformations 147

**a** Take the original graph, and shift it 2 units to the left. The result is shown in blue.

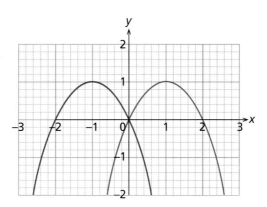

**b** Take the answer to part a, and reflect it in the x-axis. The result is shown in green.

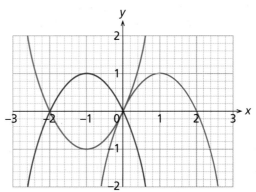

## Exercise 8.7

**You will need:**
- squared paper

**1** The graph of $y = f(x)$ is shown. Sketch the graphs of
  **a** $y = 2f(x)$
  **b** $y = -2f(x)$
  **c** $y = 2f(-x)$

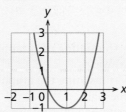

**2** The graph of $y = f(x)$ is shown. Sketch the graphs of
  **a** $y = f(2x)$
  **b** $y = -f(2x)$
  **c** $y = f(-2x)$

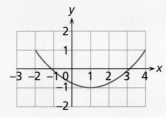

# 148 TRANSFORMATIONS OF GRAPHS

**3** The graph of $y = f(x)$ is shown.
Sketch the graph of $y = f(x + a) + b$,
where $a$ and $b$ are positive.

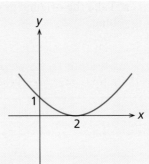

**4** The graph of $y = f(x)$ is shown.
Sketch the graph of $y = -2f(x)$.

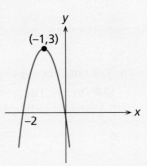

**5** The graph of $y = f(x)$ is shown.
Sketch the graph of $y = f(-2x)$.

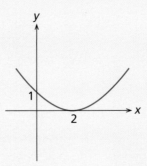

**6** Take any graph, $y = f(x)$. Sketch the graph of $y = 2f(\frac{1}{2}x)$. What single transformation has taken the first graph to the second?

## SUMMARY

- The graph of $y = f(x) + a$ is obtained by translating the graph of $y = f(x)$ by $a$ units upwards.
  If $a$ is negative, the graph is moved downwards.
- The graph of $y = f(x + a)$ is obtained by translating the graph of $y = f(x)$ by $a$ units to the left.
  If $a$ is negative, the graph is moved to the right.
- The graph of $y = af(x)$, where $a > 1$, is obtained by stretching the graph of $y = f(x)$ by a factor of $a$ in the direction of the $y$-axis.
  If $0 < a < 1$, the graph is compressed.
- The graph of $y = f(ax)$, where $a > 1$, is obtained by compressing the graph of $y = f(x)$ by a factor of $a$ in the direction of the $x$-axis.
  If $0 < a < 1$, the graph is stretched.
- The graph of $y = -f(x)$ is obtained by reflecting the graph of $y = f(x)$ in the $x$-axis.
- The graph of $y = f(-x)$ is obtained by reflecting the graph of $y = f(x)$ in the $y$-axis.
- These different transformations can be combined.

# Exercise 8A

**You will need:**
- squared paper

1 Sketch the graph of $y = x^2 + 3$.
2 The graph of $y = f(x)$ is shown. Sketch the graph of $y = f(x + 1)$, writing down the coordinates of the maximum and of the $x$-intercept.

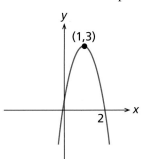

3 The graph of $y = x^2 - 2x - 3$ is shown. The points marked $A$, $B$, $C$ and $D$ are at $(-1, 0)$, $(0, -3)$, $(1, -4)$ and $(3, 0)$. Sketch the graph of $y = 3x^2 - 6x - 9$, giving the coordinates of the new positions of $A$, $B$, $C$ and $D$.

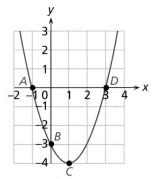

4 Refer to the graph of $y = x^2 - 2x - 3$ in question 3. Sketch the graph of $y = (2x)^2 - 2(2x) - 3$, giving the coordinates of the new positions of $A$, $B$, $C$ and $D$.
5 Sketch the graph of $y = -x^2 + 2x + 3$.
6 The graph of $y = f(x)$ is shown. Sketch the graphs of
   **a** $y = f(x - 2)$
   **b** $y = f(2 - x)$

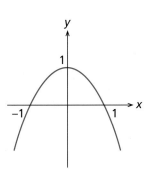

# Exercise 8B

You will need:
- squared paper

1 Sketch the graph of $y = (x-2)^2$.
2 The graph of $y = f(x)$ is shown. Sketch the graph of $y = 2f(x)$, writing down the coordinates of the maximum and of the $y$-intercept.

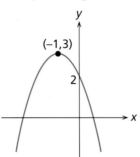

3 The graph of $y = 4x - x^2 + 12$ is shown. The points marked $A$, $B$, $C$ and $D$ are at $(-2, 0)$, $(0, 12)$, $(2, 16)$ and $(6, 0)$. Sketch the graph of $y = \frac{1}{2}(4x - x^2 + 12)$, giving the coordinates of the new positions of $A$, $B$, $C$ and $D$.

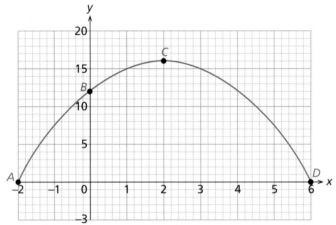

4 Refer to the graph of $y = 4x - x^2 + 12$ in question 3. Sketch the graph of $y = 4(\frac{1}{3}x) - (\frac{1}{3}x)^2 + 12$, giving the coordinates of the new positions of $A$, $B$, $C$ and $D$.
5 Sketch the graph of $y = 4(-x) - (-x)^2 + 12$.
6 The graph of $y = f(x)$ is shown. Sketch the graphs of
   a $y = -f(x)$
   b $y = 2 - f(x)$

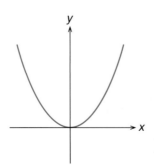

## Exercise 8C

**You will need:**
- squared paper

In this chapter we have dealt with translations, stretches and compressions, and reflections in the axes. What about other transformations?

1 Rotate the graph of $y = x^2$ by 90° anti-clockwise about the origin. What does it look like?
2 Refer to the graph of $y = 4x - x^2$ (page 138). Rotate it through 180° about the origin. What does it look like?
3 Reflect the graph of $y = 4x - x^2$ in the line $y = x$. Sketch the result.

## Exercise 8D

**You will need:**
- squared paper

1 Use the graph of $y = x^2$ to sketch the graph of $y = (-x)^2$. What do you notice? Try some other functions.
2 The graph of $y = f(x) = x^3 - 3x$ is shown. Use it to sketch the graphs of $y = f(-x)$ and then of $y = -f(-x)$. What do you notice? Try some other functions.

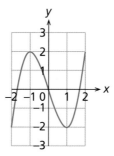

## Exercise 8E

**You will need:**
- graphics calculator
  OR
- computer with graph-drawing software installed

In this chapter you sketched graphs by hand. If you have a graphics calculator or a computer graph-drawing package then you can verify many of the results. (Notice that you only verify them: most of the results of this chapter were about a general function $y = f(x)$.) Pick any function, plot it on the graphics calculator or the computer screen, then plot the various transformations of this chapter. Do they do what they should?

For example, you might try $y = \dfrac{12}{x^2 + 2x + 4}$. Plot this, then plot

$y = \dfrac{24}{x^2 + 2x + 4}$     $y = \dfrac{12}{(x-1)^2 + 2(x-1) + 4}$

$y = \dfrac{12}{x^2 + 2x + 4} - 2$     $y = \dfrac{12}{(2x)^2 + 2(2x) + 4}$

$y = \dfrac{-12}{x^2 + 2x + 4}$     $y = \dfrac{12}{x^2 - 2x + 4}$

# 9 Trigonometry for all angles

**Remember:**
SOHCAHTOA

$\sin x = \dfrac{\text{OPP}}{\text{HYP}}$  $\cos x = \dfrac{\text{ADJ}}{\text{HYP}}$

$\tan x = \dfrac{\text{OPP}}{\text{ADJ}}$

The trigonometric functions **sine**, **cosine** and **tangent** are defined in terms of right-angled triangles. But they have many uses besides finding lengths and angles in triangles. For these uses, we often need to define sine, cosine and tangent for every angle, not just acute angles. You will find that your calculator can evaluate the trigonometric functions for angles less than 0° or greater than 90°.

## Exercise 9.1

**You will need:**
- graph paper OR
- graphics calculator OR
- graph plotter

1 Use your calculator to find the following.
   **a** $\sin 100°$  **b** $\cos 240°$  **c** $\tan -82°$

2 Plot the graph of $y = \sin x$ for $0° \leq x \leq 360°$. If you have a graphics calculator or a graph plotter on a computer then you can plot it on the screen. If you have an ordinary scientific calculator use it to fill in the table below.

| x | 0° | 30° | 60° | 90° | 120° | 150° | 180° | 210° | 240° | 270° | 300° | 330° | 360° |
|---|---|---|---|---|---|---|---|---|---|---|---|---|---|
| sin x | | | | | | | | | | | | | |

The results of question 2 above should look like this.

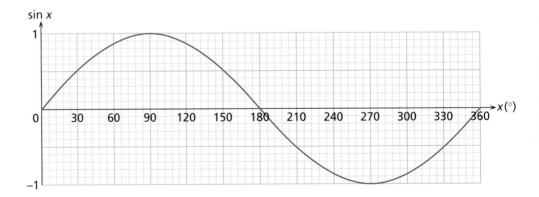

Notice that the graph goes up and down in a 'wave' shape. In fact, the sine function is used to model any quantity that goes up and down with a regular period. Some examples of such quantities are

- sound, for example the vibration caused by a musical instrument
- alternating electrical current
- the level of the sea or a tidal river.

The graph of $y = \cos x$ is similar, but starts at a different stage of the cycle.

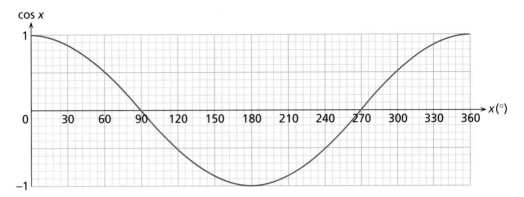

The graph of $y = \tan x$ is different. Notice that it is not defined at $x = 90°$ or $x = 270°$. If you try to find $\tan 90°$ on a calculator you will get an error message. As $x$ gets closer to $90°$, the value of $\tan x$ increases rapidly. The lines $x = 90°$ and $x = 270°$ are **asymptotes** of the graph. The curve approaches them, but never reaches them.

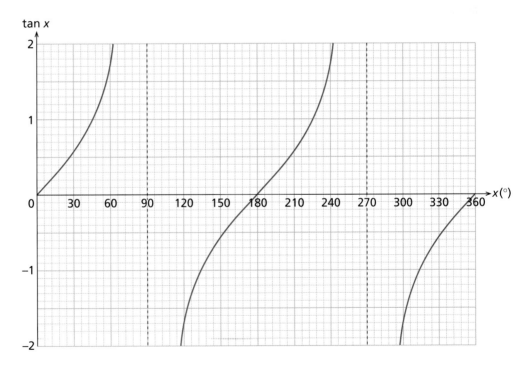

# Exercise 9.2

Use the graphs of $y = \sin x$, $y = \cos x$ and $y = \tan x$ to estimate the following. Then check your results with a calculator.

**a** $\sin 160°$  **b** $\cos 150°$  **c** $\sin 220°$  **d** $\cos 320°$
**e** $\tan 140°$  **f** $\tan 300°$

# The unit circle

We now have a look at the definition of sine and cosine, in a way which will lead on easily and naturally to angles greater then 90°.

The diagram shows a circle of radius 1 unit, with centre at the origin. A point $P$ on the circle is at $(x, y)$. Note that $x$ and $y$ must obey Pythagoras' theorem

$$x^2 + y^2 = 1^2 = 1$$

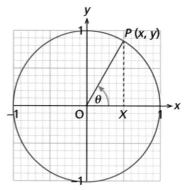

Draw the radius $OP$ making $\theta$ with the $x$-axis.

Let $X$ be the foot of the perpendicular from $P$ to the $x$-axis. Then $OX$ is the $x$-coordinate of $P$. By trigonometry

$$\cos\theta = \frac{OX}{OP} = \frac{x}{1}$$

Hence $x = \cos\theta$.

Similarly,

$$\sin\theta = \frac{PX}{OP} = \frac{y}{1}$$

Hence $y = \sin\theta$.

So the coordinates of $P$ are $(\cos\theta, \sin\theta)$.

By trigonometry

$$\tan\theta = \frac{PX}{OX} = \frac{y}{x} = \frac{\sin\theta}{\cos\theta}$$

So the tangent of any angle $\theta$, for $0° \leq \theta < 90°$, is the ratio of the sine to the cosine. For example, if $\sin\theta = 0.6$ and $\cos\theta = 0.8$ then

$$\tan\theta = \frac{\sin\theta}{\cos\theta} = \frac{0.6}{0.8} = \frac{3}{4}$$

**Note.** This equation explains why $\tan 90°$ is not defined. The cosine of 90° is 0, and we cannot divide by 0.

*Example* Show that the point $A(\tfrac{4}{5}, \tfrac{3}{5})$ lies on the unit circle. Find the angle that $OA$ makes with the $x$-axis.

Let $x = \tfrac{4}{5}$ and $y = \tfrac{3}{5}$. Check that $x^2 + y^2 = 1$.

$$(\tfrac{4}{5})^2 + (\tfrac{3}{5})^2 = \tfrac{16}{25} + \tfrac{9}{25} = \tfrac{25}{25} = 1$$

$(\tfrac{4}{5}, \tfrac{3}{5})$ lies on the unit circle.

If $OA$ makes $\theta$ with the $x$-axis, then the $y$-coordinate of $A$ is $\sin\theta$.

$$\theta = \sin^{-1}\tfrac{3}{5}$$

The angle is 36.9°.

## Exercise 9.3

1. Show that $P(0.6, 0.8)$ lies on the unit circle. Find the angle that $OP$ makes with the $x$-axis.
2. Show that $P(0.96, 0.28)$ lies on the unit circle. Find the angle that $OP$ makes with the $x$-axis.
3. If $\cos\theta = 0.6$ and $\sin\theta = 0.8$, find $\tan\theta$.
4. If $\sin\theta = \frac{8}{17}$ and $\cos\theta = \frac{15}{17}$, find $\tan\theta$.
5. If $\tan\theta = \frac{3}{4}$ and $\cos\theta = \frac{4}{5}$, find $\sin\theta$.
6. If $\tan\theta = \frac{7}{24}$ and $\cos\theta = \frac{24}{25}$, find $\sin\theta$.
7. If $\tan\theta = \frac{5}{12}$ and $\sin\theta = \frac{5}{13}$, find $\cos\theta$.
8. If $\tan\theta = \frac{40}{9}$ and $\sin\theta = \frac{40}{41}$, find $\cos\theta$.

We now extend the definition, by taking points on the unit circle in the other quadrants. In all cases, if $P$ is on the unit circle and $OP$ makes $\theta$ with the $x$-axis, then the $x$-coordinate of $P$ is $\cos\theta$ and the $y$-coordinate of $P$ is $\sin\theta$.

The $x$- and $y$-axes divide the unit circle into four **quadrants**. As $P$ goes anti-clockwise round the circle, the angle $\theta$ increases from $0°$ to $360°$.

The four quadrants are defined below. The $x$- and $y$-coordinates are sometimes negative, as is given in the table.

| | | |
|---|---|---|
| $0° < \theta < 90°$ | first quadrant | $x>0, y>0$ |
| $90° < \theta < 180°$ | second quadrant | $x<0, y>0$ |
| $180° < \theta < 270°$ | third quadrant | $x<0, y<0$ |
| $270° < \theta < 360°$ | fourth quadrant | $x>0, y<0$ |

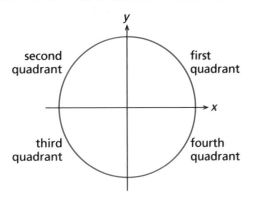

In the first quadrant, where $0° < \theta < 90°$, the point $P$ is at $(\cos\theta, \sin\theta)$. We extend the definition of sine, cosine and tangent so that, for all four quadrants, the $x$-coordinate of $P$ is $\cos\theta$ and the $y$-coordinate of $P$ is $\sin\theta$. $\tan\theta$ is $\sin\theta$ divided by $\cos\theta$.

$$\tan\theta = \frac{\sin\theta}{\cos\theta}$$

Look again at the graphs of $y = \sin\theta$, $y = \cos\theta$ and $y = \tan\theta$, for $0° \le \theta \le 360°$. Notice that

- the sine graph is negative for $180° < \theta < 360°$ (3rd and 4th quadrants)
- the cosine graph is negative for $90° < \theta < 270°$ (2nd and 3rd quadrants)
- the tangent graph is negative for $90° < \theta < 180°$ and for $270° < \theta < 360°$ (2nd and 4th quadrants).

We can use the unit circle to show how the graph of $\sin\theta$ is obtained.

# 156 TRIGONOMETRY FOR ALL ANGLES

*Example*  Plot points at every 30° round the unit circle. For each point, measure the y-coordinate. Plot the graph of the y-coordinate against the angle θ.

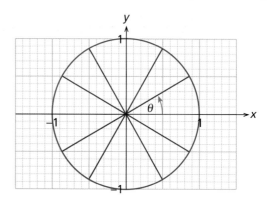

The table below gives the y-coordinates, correct to the nearest 0.05, for each value of θ.

| θ | 0° | 30° | 60° | 90° | 120° | 150° | 180° | 210° | 240° | 270° | 300° | 330° | 360° |
|---|---|---|---|---|---|---|---|---|---|---|---|---|---|
| y | 0 | 0.5 | 0.85 | 1 | 0.85 | 0.5 | 0 | −0.5 | −0.85 | −1 | −0.85 | −0.5 | 0 |

The graph of y against θ is shown. Notice that it is the same as the graph of sin x found by calculator or computer on page 152.

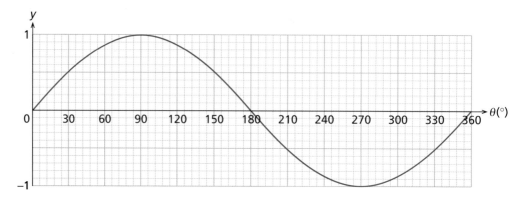

## Exercise 9.4

**You will need:**
- graph paper OR
- graphics calculator OR
- graph plotter

1 For each point on the unit circle of the example above, measure the x-coordinate. Plot the graph of the x-coordinate against θ. Is it the same as the graph of cos x on page 153?

2 For each point on the unit circle of the example, find the ratio
$$t = \frac{y\text{-coordinate}}{x\text{-coordinate}}$$
Plot a graph of t against θ. Is it the same as the graph of tan x on page 153?

# Finding angles

Suppose you want to find the angles between 0° and 360° which obey an equation of the form $\sin \theta = k$, or $\cos \theta = -k$ and so on.

Look again at the graph of $y = \sin \theta$ or of $y = \cos \theta$. If $k$ lies between $-1$ and $1$, there are two possible values of $\theta$. For example, the diagram shows the graph of $y = \sin \theta$. Notice that the line $y = 0.5$ crosses the graph twice, once in the first quadrant and once in the second. Hence there are two solutions to the equation $\sin \theta = 0.5$.

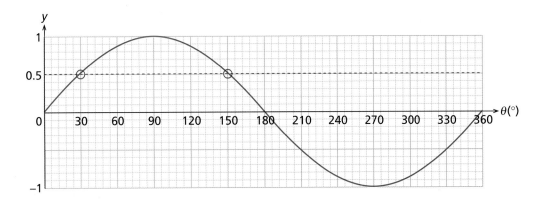

The first solution is in the first quadrant, between 0° and 90°. This solution is $\sin^{-1} 0.5$, which is 30°.

The next solution is in the second quadrant, between 90° and 180°. We can use the symmetry of the graph to find the second solution. The first solution is 30° to the right of 0°, so the second solution is 30° to the left of 180°.

second solution = 180° − 30° = 150°

You can use your calculator to check that $\sin 150° = 0.5$.

Now suppose we want to solve $\sin x = -0.5$. The line $y = -0.5$ crosses the sine curve twice, as shown below.

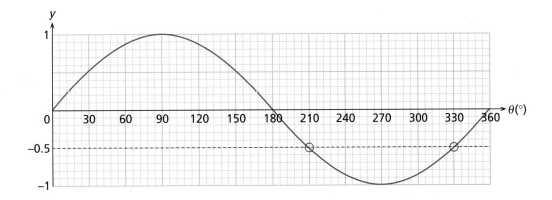

## 158 TRIGONOMETRY FOR ALL ANGLES

These solutions are in the second half of the graph, which is similar to the first half, but reflected in the *x*-axis. So the solutions for $\sin x = -0.5$ can be found from those for $\sin x = +0.5$, by adding on 180°.

$$30° + 180° = 210° \qquad (180° - 30°) + 180° = 360° - 30° = 330°$$

> If you try to find $\sin^{-1}(-0.5)$ directly from your calculator, you will get $-30°$, which is outside the range 0° to 360°.

You can check with your calculator that $\sin 210° = \sin 330° = -0.5$.

So the method of solving such equations is as follows. Use the graph to identify the quadrants in which there are solutions. Use your calculator to find the angle whose sine has a positive value. Then, as appropriate

| First quadrant | give the angle directly |
| Second quadrant | subtract the angle from 180° |
| Third quadrant | add 180° to the angle |
| Fourth quadrant | subtract the angle from 360° |

The same method applies to the cosine and tangent functions.

*Example* Find the angles between 0° and 360° for which
  **a** $\sin \theta = 0.8$  **b** $\cos \theta = -0.46$  **c** $\tan \theta = -1.5$

**a** From the graph, we see that sine is positive in the first and second quadrant.

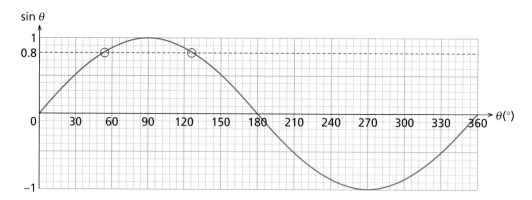

For the value in the first quadrant, use a calculator to find $\sin^{-1} 0.8 = 53.1°$.
For the value in the second quadrant, subtract this from 180°.

$$180° - 53.1° = 126.9°$$

$\theta = 53.1°$ or $126.9°$.

**b** From the graph, we see that cosine is negative in the second and third quadrant.

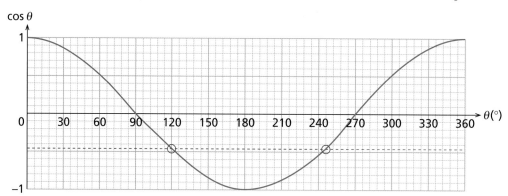

First find the angle whose cosine is +0.46, as 62.6°. For the value in the second quadrant, subtract from 180°. For the value in the third quadrant, add 180°.

$$\cos^{-1} 0.46 = 62.6°$$
$$180° - 62.6° = 117.4° \qquad 180° + 62.6° = 242.6°$$

$\theta = 117.4°$ or $242.6°$.

c From the graph, tangent is negative in the second and fourth quadrant.

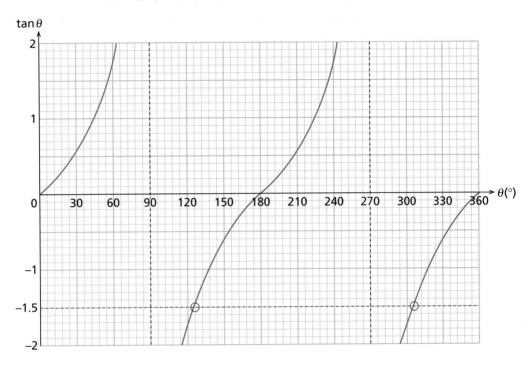

First find the angle whose tangent is +1.5. For the value in the second quadrant, subtract from 180°. For the value in the fourth quadrant, subtract from 360°.

$$\tan^{-1} 1.5 = 56.3°$$
$$180° - 56.3° = 123.7° \qquad 360° - 56.3° = 303.7°$$

$\theta = 123.7°$ or $303.7°$.

**Note.** Though your calculator will not give all these values directly, you can still use it to *check* that the answer is right. Check, for example, that $\tan 303.7° = -1.5$.

# TRIGONOMETRY FOR ALL ANGLES

## Exercise 9.5

For each of the equations in questions 1–12, find two solutions between 0° and 360°.

1. $\sin x = 0.5$
2. $\cos x = -0.5$
3. $\cos x = 0.5$
4. $\sin x = -0.5$
5. $\tan x = 1$
6. $\tan x = -1$
7. $\cos x = 0.48$
8. $\cos x = -0.94$
9. $\sin x = 0.55$
10. $\sin x = -0.35$
11. $\tan x = -0.42$
12. $\tan x = 2.3$
13. Suppose $\sin x = 0.7$. Find $x$, and hence find two possible values of $\cos x$.
14. Suppose $\cos x = 0.2$. Find $x$, and hence find two possible values of $\sin x$.
15. Suppose $\tan x = 2$. Find $x$, and hence find two possible values of $\cos x$ and $\sin x$.
16. Suppose $\tan x = -0.3$. Find $x$, and hence find two possible values of $\cos x$ and $\sin x$.
17. Solve the equation $\sin x = 0.6$, for values between 0° and 360°. Hence find the values of $x$ in this range for which $\sin x > 0.6$.
18. Solve the following inequalities in the range 0° to 360°.
    a. $\sin x < -0.5$
    b. $\cos x > 0.5$
    c. $\cos x < -0.3$
    d. $\tan x > 1$
    e. $\tan x < 1$
    f. $\tan x > -1$

# Negative angles

An anti-clockwise rotation is positive. When $P$ went round the unit circle, the radius $OP$ rotated anti-clockwise about the origin. If $OP$ rotates clockwise, then the angle between $OP$ and the $x$-axis takes negative values. In the diagram, the value of $\theta$ is negative, so the radius $OP$ lies below the $x$-axis.

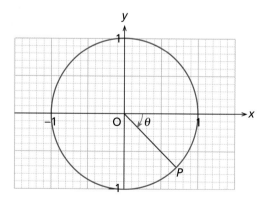

Notice that $x > 0$ and $y < 0$. Hence $\cos \theta$ remains positive, $\sin \theta$ becomes negative. As $\tan \theta$ is a negative number divided by a positive number, $\tan \theta$ is negative.

$$\cos -\theta = \cos \theta \qquad \sin -\theta = -\sin \theta \qquad \tan -\theta = -\tan \theta$$

# Negative angles

We can extend the graphs of sine, cosine and tangent for negative values of the angle. The graphs are shown for angles between −360° and 360°.

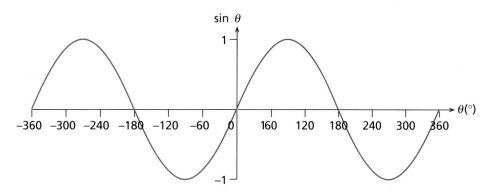

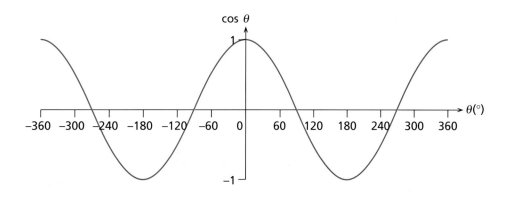

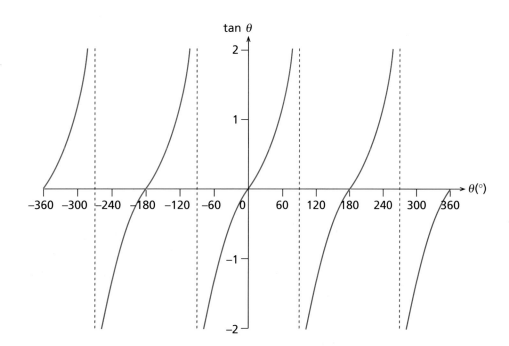

## Angles greater than 360°

If θ is greater than 360°, then the radius $OP$ has rotated through more than a full circle. The *x*- and *y*-coordinates repeat themselves, after each rotation of 360°.

These diagrams show the graphs of $y = \sin \theta$, $y = \cos \theta$ and $\tan \theta$, for angles up to $2 \times 360°$, which is 720°. Notice that the graphs between 360° and 720° are copies of the graphs between 0° and 360°.

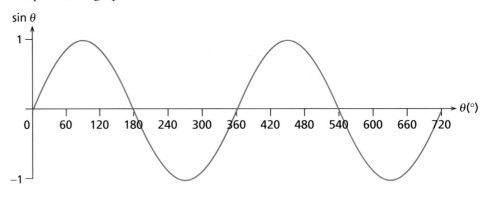

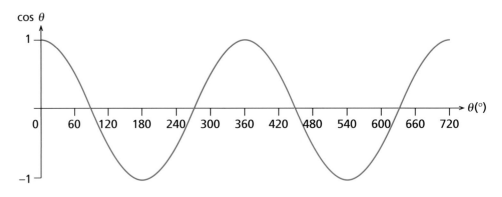

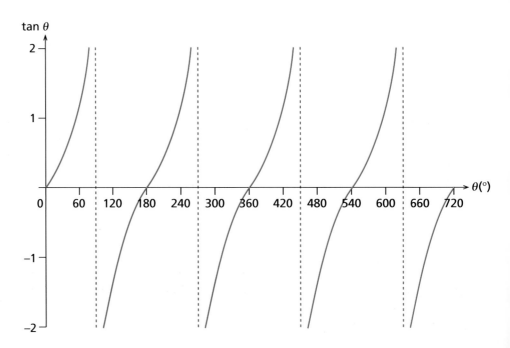

# Transformations of trigonometric graphs

The trigonometric graphs can be transformed by the operations of chapter 8. This is particularly important for applications of the sine or cosine graph, as we shall see. Recall from the previous chapter that if we start with the graph of $y = f(x)$:

- the graph of $y = af(x)$ is obtained by stretching the graph in the direction of the $y$-axis by a factor of $a$
- the graph of $y = f(ax)$ is obtained by compressing the graph in the direction of the $x$-axis by a factor of $a$
- the graph of $y = f(x + a)$ is obtained by shifting the graph by $a$ units to the left
- the graph of $y = f(x) + a$ is obtained by shifting the graph by $a$ units upwards.

Take the original graph of $y = \sin x$. The graph of $y = 3\sin x$ is obtained by stretching it by a factor of 3 in the direction of the $y$-axis, as shown. The graph now goes between $-3$ and $+3$.

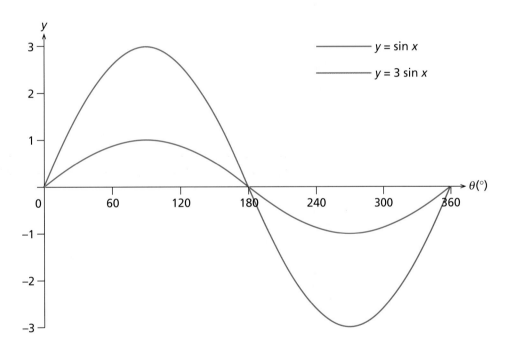

> Suppose the graph of $y = \sin x$ represents a sound wave of the note produced by a musical instrument. Then the graph of $y = 3\sin x$ will be the same note, but much louder. The graph of $y = \sin 2x$ will be a note of twice the frequency, i.e. an octave higher.

The graph of $y = \sin 2x$ is obtained by compressing by a factor of 2 in the direction of the $x$-axis, as shown. The graph still goes between $-1$ and $+1$, but it happens twice as fast. The graph repeats itself after every $180°$, rather than every $360°$.

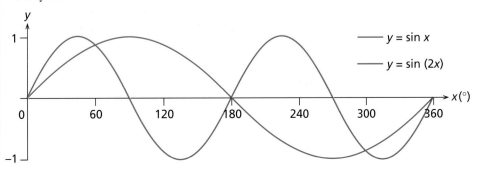

The graph of $y = \sin(x + 30°)$ is obtained by shifting the original graph 30° to the left. The result is shown.

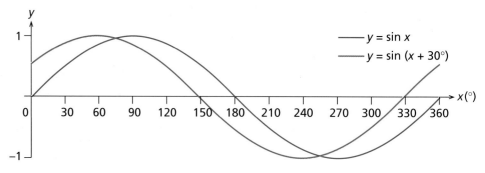

The graph of $y = \sin x + 1$ is obtained by shifting the original graph upwards. The result is shown. Notice that the graph now goes between 0 and 2.

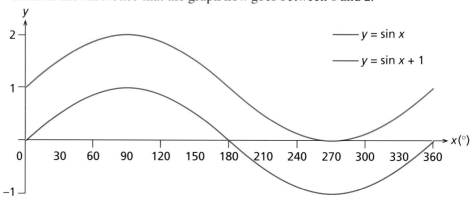

*Examples*  Sketch the graph of $y = 2\sin 3x$, for $0° \leq x \leq 360°$.

Two transformations are happening here. Because of the factor of 2, the graph goes between −2 and +2. Because of the factor of 3, the graph repeats itself every $360° \div 3$, i.e. every 120°. The graph is shown.

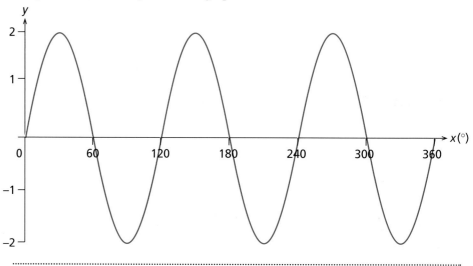

At a harbour, the depth of the sea varies throughout the day. On a particular day, at $t$ hours after midnight, the depth $d$ m is given by

$$d = 5 + 2 \cos 30t$$

**a** Sketch a graph of $d$ against $t$, for a 24-hour period.
**b** A ship can dock at the harbour provided the depth of water is at least 6 m. When in the morning can the ship dock?

**a** As $t$ goes from 0 to 24, $30t$ goes from $0°$ to $720°$. So there are two complete cycles of the cosine graph.
The value of $2 \cos 30t$ goes between $-2$ and $+2$, so $d$ goes between 3 and 7. The graph is shown.

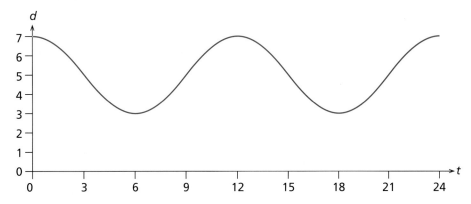

**b** We need $d \geq 6$. So we need

$$5 + 2 \cos 30t \geq 6$$
$$\cos 30t \geq \tfrac{1}{2}$$

The time is in the morning, so $t \leq 12$. From the methods of the previous section, $\cos 30t = \tfrac{1}{2}$ when $30t = 60°$ or $300°$. This gives $t = 2$ or $t = 10$. Looking at the graph, we want the period before 2 or after 10.

The ship can dock up to 2 a.m. or after 10 a.m.

## Exercise 9.6

You will need:
• graph paper

**1** Sketch the following graphs, for $0° \leq x \leq 360°$.
  **a** $y = 2 \sin x$
  **b** $y = 3 \cos x$
  **c** $y = \sin 3x$
  **d** $y = \cos 2x$
  **e** $y = 0.5 \cos x$
  **f** $y = \tfrac{1}{3} \sin x$
  **g** $y = \sin \tfrac{1}{2} x$
  **h** $y = \cos \tfrac{1}{3} x$
  **i** $y = 2 \cos 3x$
  **j** $y = 4 \sin 2x$
  **k** $y = 2 \sin \tfrac{1}{2} x$
  **l** $y = 3 \cos \tfrac{1}{2} x$
  **m** $y = \sin(x - 30°)$
  **n** $y = \cos(x + 60°)$
  **o** $y = \sin(x + 90°)$
  **p** $y = \cos(x - 90°)$
  **q** $y = \sin x + 2$
  **r** $y = \cos x - 1$

**2** The graphs below are all of the form $y = k \sin mx$. In each case find $k$ and $m$.

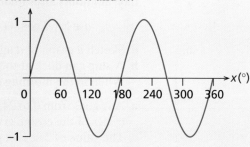

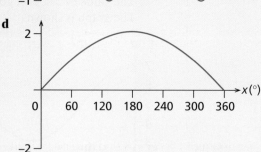

**3** At $t$ hours after midnight, the depth of water in a harbour is $d$ metres, where $d = 5 + 4 \sin 30t$.
  **a** Sketch the graph of $d$ against $t$.
  **b** When is high tide?
  **c** When is low tide?
  **d** A ship can dock at the harbour provided that the depth is at least 7 m. When can the ship dock?

**4** On the $d$th day after midsummer (i.e. $d$ days after 21 June) the length of the day is $l$ hours, where $l$ is approximately $12 + 2 \cos d$.
  **a** Sketch the graph of $l$ against $d$.
  **b** According to this model, when is midwinter? When are the equinoxes?
  **c** When is there more than 13 hours of daylight?
  **d** Give one reason why this formula is only approximate. How could it be made more accurate?

**5** A model for temperature variation throughout the day is that, $h$ hours after 3 a.m., the temperature is $T\,°C$, where $T = 20 - 10 \cos 15h$.
  **a** Plot the graph of $T$ against $h$.
  **b** What is the temperature at midday?
  **c** What is the hottest time of day?
  **d** When is the temperature below 12 °C?

# SUMMARY

- The definitions of **sine**, **cosine** and **tangent** can be extended to all angles.
- The graphs of sine and cosine have a wave-like shape.
- Sine and cosine curves occur naturally in waves such as sound waves and water waves.
- The graph of tangent has **asymptotes** at 90°, 270°, etc.
- In the second **quadrant**, sine is positive and the other two functions are negative. In the third quadrant, only tangent is positive and in the fourth quadrant only cosine is positive.
- The sine or cosine graphs can be transformed by stretching the $y$-values or by compressing the $x$-values. The graph of $y = a \sin bx$ is stretched by a factor of $a$ in the direction of the $y$-axis, and compressed by a factor of $b$ in the direction of the $x$-axis.

## Exercise 9A

**You will need:**
- graph paper

1. Refer to this graph of $y = \sin x$. Use it to find the approximate value of $\sin 240°$.

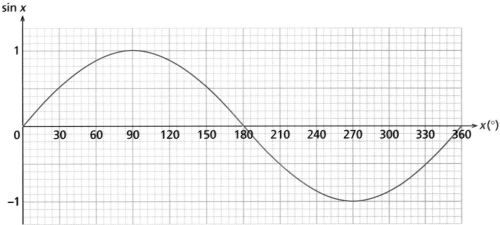

2. Use the graph of $y = \sin x$ to find the approximate solutions of $\sin x = -0.7$.
3. Solve, for $0° \leq x \leq 360°$, $\cos x = 0.4$.
4. Solve, for $0° \leq x \leq 360°$, $\sin x = -0.7$.
5. Sketch the graph of $y = \cos x$ for $-180° \leq x \leq 180°$.
6. Sketch the graph of $y = 5\cos 2x$ for $0° \leq x \leq 360°$.
7. Sketch the graph of $y = \sin(x + 50°)$ for $0° \leq x \leq 360°$.
8. The 'London Eye', the big wheel on the south bank of the Thames, rotates so that, $t$ minutes after starting, the height of a passenger above the ground is $h$ metres, where $h = 70 - 70\cos 12t$.
   a. What is the greatest height a passenger reaches?
   b. A ride consists of one circuit. How long does a ride last?
   c. When is the passenger 25 m high?

## Exercise 9B

**You will need:**
- graph paper

1. Refer to this graph of $y = \cos x$. Use it to find the approximate value of $\cos 160°$.

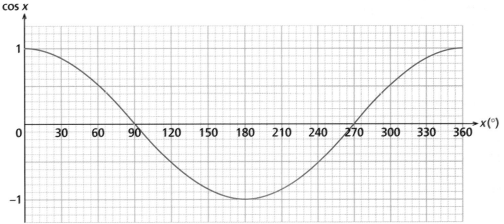

2. Use the graph of $y = \cos x$ to find the approximate solutions of $\cos x = 0.3$.
3. Solve, for $0° \leq x \leq 360°$, $\tan x = -1.4$.
4. Solve, for $0° \leq x \leq 360°$, $\cos x = 0.3$.

5 Sketch the graph of $y = \tan x$ for $0° \leq x \leq 180°$
6 Sketch the graph of $y = \frac{1}{2}\sin 3x$, for $0° \leq x \leq 360°$.
7 Sketch the graph of $y = 2 + \sin x$, for $0° \leq x \leq 360°$.
8 The demand for electricity varies throughout the day. If $t$ is the time in hours after 6 a.m., the demand in a town is $d$ gigawatts, where $d = 6 + 3\sin 15t$.
  a Sketch a graph of $d$ against $t$.
  b What is the greatest demand, and when does it occur?
  c When is the demand greater than 7 gigawatts?

## Exercise 9C

You have used trigonometry to solve many problems involving bearings. You adjusted the bearing so that you were dealing with an acute angle. We have now defined trigonometric functions for all angles, so we can use them directly for any bearing.

1 A ship sails 100 km on a bearing of 055°. Verify that it has gone 100 cos 55° km north and 100 sin 55° km east.
2 A ship sails 100 km on a bearing of 155°. Verify that it has gone 100 cos 155° km north and 100 sin 155° km east.

> A negative distance north is the same as a positive distance south.

3 Try similar calculations in the other two quadrants. Verify that, for any bearing $b$, after sailing 100 km on a bearing of $b$, the ship has gone 100 cos $b$ km north and 100 sin $b$ km east.

## Exercise 9D

You will need:
- graph paper OR
- graphics calculator OR
- graph plotter

Some of the exercises at the end of this chapter involved using a sine or cosine function for a quantity such as temperature, sea level or length of day. Is the function a good fit? Collect data, draw a graph and see whether it fits a sine or a cosine function. You could try the examples of this chapter, for example:

1 the temperature over the day
2 the level of the sea over the day (if you live near the sea you can use tide tables)
3 the length of the day over the year (a diary or newspaper may have this information).

## Exercise 9E

You will need:
- graphics calculator OR
- spreadsheet OR
- graph plotter

The sine and cosine curves can be used to model the sound waves produced by musical instruments. The sound wave of a flute, for example, is very close to a sine curve. For other instruments, in particular for metal instruments such as a trumpet or a harmonica, the sound wave is built up of the basic note with many overtones, which are notes of one or more octaves higher. In this exercise you see what the curve of a note with overtones looks like.

1 Plot the graph of $y = \sin x + \frac{1}{2}\sin 2x$. What does it look like?
2 Let $y = \sin x + \frac{1}{3}\sin 3x + \frac{1}{5}\sin 5x + \cdots$ (continue for as many terms as you like). Plot this graph. What does it look like?
3 Try some other combinations of sine functions, and plot them.

# 10 Sequences

A **sequence** is any succession of numbers following a rule. The numbers could be corresponding to patterns or shapes, or they could be purely numerical. Here are some examples of sequences.

1, 3, 5, 7, ...   the sequence of odd numbers
1, 4, 9, 16, ...   the sequence of square numbers
1, 3, 6, 10, ...   the sequence corresponding to the patterns of dots below.

> These numbers are called **triangular** numbers.

There are two ways to describe a sequence.

- Give the first term and the rule which describes how to go from one term to the next.
- Give a formula which gives the general term, the $n$th term.

For example, consider the sequence of odd numbers above. The two descriptions could be as follows.

- The first term is 1. To go from one term to the next, add 2.
- The $n$th term of the sequence is $2n - 1$.

It is often quicker to use the formula for the general term. Suppose we want to find the 100th odd number. It would be very tedious to start with 1, and then add 2 ninety-nine times! With the formula we can go directly to the 100th term, without having to find all its predecessors.

The 100th odd number is $2 \times 100 - 1$, which is 199.

Unfortunately it is not always possible to find a general formula for the $n$th term. In this case we have to use the first method. A computer is useful for this.

**Note.** If you are given just the first few terms of a sequence, it is not possible to determine its rule or formula. Consider the sequence which begins 1, 2, 4, .... There are many ways it could continue.

1, 2, 4, 8, 16, ...   (doubling each time)
1, 2, 4, 7, 11, ...   (difference between terms increasing by 1 each time)

## Linear sequences

Consider these sequences.

$$3, 7, 11, 15, \ldots \quad [1]$$
$$8, 7.5, 7, 6.5, \ldots \quad [2]$$

In both of these sequences there is a constant difference (known as the **common difference**) between terms. The first sequence is increasing by 4 each time, and the second is decreasing by 0.5. They are called **linear sequences**, because if you plot them as points, for example (1, 3), (2, 7), (3, 11), (4, 15), etc. for sequence [1] above, they will lie on a straight line, as shown.

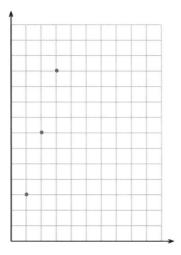

> These sequences are also known as **arithmetic** sequences.

It is always possible to find a formula for the general term of a linear sequence.
Find the common difference between terms. Say it is $d$.

For the sequence [1], $d = 4$.

Find the difference between the first term $a$ and $d$. Say the first term is $k$ greater than $d$, i.e. $a = d + k$. (If the first term is less than $d$, then $k$ is negative.)

For the sequence [1], $a = 3$. Hence $k = 3 - 4 = -1$.

Then the $n$th term is $dn + k$.

For the sequence [1], the $n$th term is $4n - 1$.

Here is another way to find the $n$th term.

| Say the first term is $a$. | For [1], $a = 3$ |
| The second term is $a + d$. | $a + d = 7$ |
| The third term is $a + 2d$ | $a + 2d = 11$ |

And so on. At the $n$th term, $d$ has been added on $(n - 1)$ times. (It is $n - 1$, not $n$.) Hence the $n$th term is

$$a + (n - 1)d$$

For the sequence [1], the $n$th term is $3 + (n - 1)4 = 3 + 4n - 4 = -1 + 4n$.

**Note.** There is much in common between a linear sequence and a linear graph with equation $y = mx + c$. The first term of the sequence corresponds to the value of $c$, the $y$-intercept. The common difference of the sequence corresponds to the value of $m$, the gradient of the graph.

We can always check the result. Try the third term of sequence [1], i.e. put $n = 3$ into the formula $4n - 1$.

$$4 \times 3 - 1 = 11$$

This is correct.

*Example* Find a formula for the $n$th term of the sequence [2] above.

The sequence is 8, 7.5, 7, 6.5, .... The common difference $d$ is $-0.5$. The first term is 8, which is 8.5 greater than $-0.5$. Hence $k = 8.5$. We can now write down the formula.
The $n$th term is $-0.5n + 8.5$.

Alternatively, the first term is 8 and the common difference is $-0.5$. At the $n$th term, $-0.5$ has been added, $(n - 1)$ times. So the $n$th term is

$$8 + (n - 1)(-0.5)$$
$$= 8 - 0.5n + 0.5$$
$$= 8.5 - 0.5n$$

This is the same answer as above.
Check: try the fourth term. $-0.5 \times 4 + 8.5 = -2 + 8.5 = 6.5$. This is correct.

# Exercise 10.1

**1** Find a formula for the $n$th term of each of the following sequences.
  **a** 1, 4, 7, 10, ...
  **b** 3, 8, 13, 18, ...
  **c** 2, 2.5, 3, 3.5, ...
  **d** 7, 5, 3, 1, ...
  **e** 4, −1, −6, −11, ...

**2** Draw the next shape in the sequence below. How many dots are there in the $n$th shape?

**3** Draw the next shape in the sequence below. How many lines are there in the $n$th shape?

**4** Draw the next shape in the sequence below. How many squares are there in the $n$th shape?

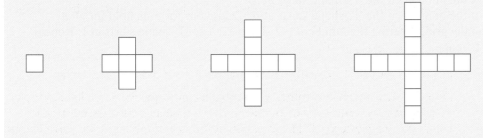

**5** There is a new moon every 29 days. Suppose there is a new moon on 6 January. On what day of the year (not a leap year) is the $n$th new moon ($n \leq 12$)?

**6** If there is a lunar eclipse, then about 18 years later there will be another lunar eclipse. If there is a lunar eclipse in March 2006, when will be the $n$th lunar eclipse after that?

> The time gap between eclipses is called the Saros period, and has been known about for thousands of years. It lasts 6585.3 days, or 18 years, 11 days and 8 hours.

**7** A cat starts the week with 700 grams of cat food, and eats 50 grams per day. How much is left after $n$ days?

**8** A cup initially contains $50\,\text{cm}^3$ of water, and every second a drip of $0.03\,\text{cm}^3$ leaks out of it. How much is left after $n$ seconds?

**9** I borrow £2000, and pay the debt off at £120 per month. How much do I still owe after $n$ months?

## Exercise 10.2

Below are the numbers 1 to 100, written in a square grid.

| 1 | 2 | 3 | 4 | 5 | 6 | 7 | 8 | 9 | 10 |
|---|---|---|---|---|---|---|---|---|---|
| 11 | 12 | 13 | 14 | 15 | 16 | 17 | 18 | 19 | 20 |
| 21 | 22 | 23 | 24 | 25 | 26 | 27 | 28 | 29 | 30 |
| 31 | 32 | 33 | 34 | 35 | 36 | 37 | 38 | 39 | 40 |
| 41 | 42 | 43 | 44 | 45 | 46 | 47 | 48 | 49 | 50 |
| 51 | 52 | 53 | 54 | 55 | 56 | 57 | 58 | 59 | 60 |
| 61 | 62 | 63 | 64 | 65 | 66 | 67 | 68 | 69 | 70 |
| 71 | 72 | 73 | 74 | 75 | 76 | 77 | 78 | 79 | 80 |
| 81 | 82 | 83 | 84 | 85 | 86 | 87 | 88 | 89 | 90 |
| 91 | 92 | 93 | 94 | 95 | 96 | 97 | 98 | 99 | 100 |

**1** An 'L' shape is put on the grid as shown, over the numbers 1, 11 and 12. We say the 'L' shape starts at 1.
   **a** What is the sum of the numbers under the 'L' shape?
   **b** Move the 'L' shape one to the right, so that it starts at 2. What is the sum of the numbers?
   **c** What is the sum of the numbers if the 'L' shape starts at 3?
   **d** What is the sum of the numbers if the 'L' shape starts at $n$? Does your formula hold for all $n$?

**2** A 'T' shape is put on the grid, covering the numbers 1, 2, 3 and 12. The 'T' shape starts at 1. Repeat question 1 for the 'T' shape.

## Exponential sequences

When something is increasing or decreasing very rapidly, we often say it is exponentially increasing or decreasing. You may have read statements like the following.

- The number of cases of 'flu is increasing exponentially.
- The area of the rain forest in South America is decreasing exponentially.

In mathematics, each term of an **exponential sequence** is a constant multiple of the previous term. These sequences are also called **geometric sequences**.

Consider these sequences

$$2, 6, 18, 54, 162, \ldots \quad [1]$$
$$2, 1, \tfrac{1}{2}, \tfrac{1}{4}, \tfrac{1}{8}, \ldots \quad [2]$$

In these sequences each term is obtained by multiplying its predecessor by a fixed number. For the first example, each term is three times the one before, and for the second example each term is half the one before.

So in these sequences there is a first term $a$, and a fixed ratio $r$ between each term and its predecessor. The ratio is called the **common ratio**.

Suppose a sequence has first term $a$ and common ratio $r$.

For the sequence [1] above, $a = 2$ and $r = 3$.

The second term is $a \times r$.

For the sequence [1], the second term is $2 \times 3 = 6$.

The third term is $a \times r \times r = ar^2$.

For the sequence [1], the third term is $2 \times 3^2 = 18$.

And so on. By the $n$th term, $a$ has been multiplied by $r$, $(n-1)$ times ($n-1$ times, not $n$ times). Hence the $n$th term of the sequence is $a \times r \times r \times \ldots \times r = ar^{n-1}$.

For the sequence [1], the $n$th term is $2 \times 3^{n-1}$.

A natural example of one of these sequences is when a sum of money is invested at a fixed rate of compound interest. Suppose £1000 is invested at 7% compound interest. Then every year the amount of money is multiplied by $\frac{107}{100}$, i.e. by 1.07. The amounts of money at the beginning of each year are

£1000, £1070, £1144.90, £1225.04, ... each term being 1.07 of its predecessor.

The $n$th term, the amount at the beginning of the $n$th year, is £1000 $\times 1.07^{n-1}$.

*Example* Find the tenth term of the sequence 100, 90, 81, 72.9, ...

Each term is 0.9 of its predecessor. Hence the common ratio is 0.9. The first term is 100, and hence the $n$th term is

$$100 \times 0.9^{10-1} = 100 \times 0.9^9$$

The tenth term is 38.7.

# Exercise 10.3

1. Find the next two terms of each of the sequences below.
   a. 3, 6, 12, 24, ...
   b. 2, 10, 50, 250, ...
   c. 5, 10, 20, 40, ...
   d. 12, 6, 3, 1.5, ...
   e. 54, 18, 6, 2, ...
   f. 3, −6, 12, −24, ...
   g. 3, $1\frac{1}{2}$, $\frac{3}{4}$, $\frac{3}{8}$, ...
2. For each of the sequences in question 1, write down the first term and the common ratio.
3. Find the $n$th term for each of the sequences in question 1.
4. Find the number of rectangles in the $n$th diagram of the sequence below.

5. Find the number of lines in the $n$th diagram of the sequence below.

6. £2000 is invested at 8% compound interest. How much is there after $n$ years?
7. A radioactive material decays, losing 20% of its mass each year. If there is 0.3 kg originally, how much is left after $n$ years?
8. A fair die is rolled until it comes up 6. Find the probability that this takes $n$ goes.
9. The following is known as the paradox of 'Achilles and the tortoise'. Achilles and the tortoise have a race, and as Achilles can run 10 times as fast as the tortoise, the tortoise is given a start of 100 paces. By the time that Achilles has made up the 100 paces, the tortoise is 10 paces ahead. By the time that Achilles has made up the 10 paces, the tortoise is 1 pace ahead. And so on – Achilles will never catch up with the tortoise!

   Write down the distances Achilles has to run to catch up, and add them together. When will Achilles catch up?

> When Achilles was born, his mother tried to make him immortal by dipping him in the magical River Styx. She held him by one heel to do this, and so the heel that never got wet remained mortal and vulnerable. To this day we refer to someone's weak point as their 'Achilles heel'.

# Quadratic sequences

One of the sequences mentioned at the beginning of the chapter was the sequence of squares: 1, 4, 9, 16, ... Suppose we find the differences between the terms.

> **Remember:**
> Method of differences

$$\begin{array}{ccccccccc} 1 & & 4 & & 9 & & 16 & & 25 \\ & 3 & & 5 & & 7 & & 9 & \end{array}$$

These differences are not constant, so the sequence is not linear. But the differences themselves *do* form a linear sequence.

$$\begin{array}{ccccccccc} 1 & & 4 & & 9 & & 16 & & 25 \\ & 3 & & 5 & & 7 & & 9 & \\ & & 2 & & 2 & & 2 & & \end{array}$$

So the differences between the differences, or the second differences, are constant. A sequence for which this holds is a **quadratic sequence**.

Consider this sequence

$$2 \quad 3 \quad 6 \quad 11 \quad 18 \quad \ldots$$

The differences between successive terms are not constant. But if we write down the differences and the second differences we get

$$\begin{array}{ccccccccc} 2 & & 3 & & 6 & & 11 & & 18 \\ & 1 & & 3 & & 5 & & 7 & \\ & & 2 & & 2 & & 2 & & \end{array}$$

The second differences are constant, so the differences themselves are increasing at a constant rate. The next difference is 9, and so the next term of the original sequence is $18 + 9$, which is 27.

The $n$th term of a sequence like this is a quadratic function of $n$, i.e. it is of the form $an^2 + bn + c$, where $a$, $b$ and $c$ are constant. Here is how to find the values of $a$, $b$ and $c$.

In such a sequence, the second differences are constant. Say the second differences are each $k$. Then $a = \frac{1}{2}k$. For the sequence above, the differences are increasing by 2. Hence $a = \frac{1}{2} \times 2 = 1$.

Write out the sequence with general term $an^2$. For the sequence above, this is

$$1 \quad 4 \quad 9 \quad 16 \quad 25$$

Now subtract the $an^2$ sequence from the original sequence. We get

$$1 \quad -1 \quad -3 \quad -5 \quad -7 \quad [*]$$

This is now a linear sequence. Use the methods of the first part of this chapter to find the general term for this linear sequence. This is $bn + c$. Add this to $an^2$ and we have the general term for the quadratic sequence. For the linear sequence [*] above the $n$th term is $-2n + 3$. Hence the formula for the $n$th term of the original sequence is $n^2 - 2n + 3$.

Check: put $n = 5$. Then $5^2 - 2 \times 5 + 3 = 25 - 10 + 3 = 18$. This is correct.

Here is another method of finding $b$ and $c$. Having found that $a = 1$, the general term of the sequence is $n^2 + bn + c$. Use the fact that for $n = 1$ this term is 2 (the first term of the sequence), and for $n = 2$ this term is 3 (the second term of the sequence).

$$1^2 + b + c = 2$$
$$2^2 + 2b + c = 3$$

So we have the simultaneous equations: $b + c = 1$ and $2b + c = -1$. Solve to find that $b = -2$ and $c = 3$.

*Example* Find a formula for the $n$th term of the sequence 2, 3, 8, 17, 30, ....

Write out the sequence, its differences and the second differences.

$$\begin{array}{ccccccccc} 2 & & 3 & & 8 & & 17 & & 30 \\ & 1 & & 5 & & 9 & & 13 & \\ & & 4 & & 4 & & 4 & & \end{array}$$

Notice that the second differences are each 4. We put $a = \frac{1}{2} \times 4$, i.e. $a = 2$, and write out the sequence corresponding to $an^2$, i.e. to $2n^2$.

$$2 \quad 8 \quad 18 \quad 32 \quad 50$$

Write out the original sequence underneath and subtract from this.

$$\begin{array}{ccccc} 2 & 3 & 8 & 17 & 30 \\ 0 & -5 & -10 & -15 & -20 \end{array}$$

This is now a linear sequence. The general term is $-5n + 5$. Add this to $2n^2$. The $n$th term of the sequence is $2n^2 - 5n + 5$.
Check: put $n = 5$. Then $2 \times 5^2 - 5 \times 5 + 5 = 30$. This is correct.

## Exercise 10.4

**1** For each of the sequences below, find the next two terms.
   **a** 0, 3, 8, 15, ...
   **b** 2, 6, 12, 20, ...
   **c** 3, 9, 19, 33, ...
   **d** 1, 6, 15, 28, ...
   **e** 0, −2, −6, −12, ...
**2** For each of the sequences in question 1, find the constant second difference.
**3** Find the $n$th term of each of the sequences of question 1.
**4** Find the number of dots in the $n$th triangle below.

**5** The diagram below shows rectangles built up from dots. How many dots are there in the $n$th diagram?

**6** When a group of people meet, everyone has to shake hands with everyone else.

With 1 person there are 0 handshakes (you can't shake hands with yourself).
With 2 people there is 1 handshake.
With 3 people there are 3 handshakes (A with B, B with C, C with A).

   **a** Find the number of handshakes with 4 people.
   **b** Find the number of handshakes with 5 people.
   **c** Find the number of handshakes with $n$ people.

7 The diagram shows successive stages of building a house of cards. How many cards are needed to build a house which is *n* storeys high?

8 In the early seventeenth century the Italian scientist Galileo Galilei investigated the motion of falling objects. (He may have dropped them from the Leaning Tower of Pisa.) He proposed the law that the distances covered in successive seconds are in the ratio $1:3:5:\ldots$. Suppose that the body falls 1 unit in the first second.
   a What distance does it fall in the *n*th second?
   b What is the total distance it falls in *n* seconds?

Galileo Galilei (1564–1642) studied medicine at the University of Pisa but his real interest was always in mathematics and natural philosophy. His controversial views on the motion of the Earth led to him being suspected of heresy and eventually condemned to house arrest for life.

9 The diagrams below show lines crossing. No lines are parallel, and no three lines go through the same point.

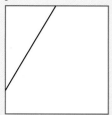 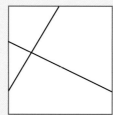 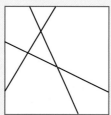

   a Draw the next diagram.
   b Write down the numbers of crossing points in each diagram. Continue this sequence for two more terms.
   c With *n* lines, how many crossing points are there?
10 For the diagrams of question 9, write down the numbers of regions. Continue this series for two more terms. With *n* lines, how many regions are there?
11 Points are put on a circle. Every possible chord joining pairs of points is drawn. Below are the diagrams for 2, 3 and 4 points. How many chords are there for *n* points?

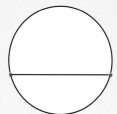  

## Exercise 10.5

**You will need:**
- computer with spreadsheet package installed

Another method of finding the $n$th term of a sequence is by trial and error, using a spreadsheet. Suppose we have the sequence 2, 3, 8, 17, 30, ...

Put 1, 2, 3, 4, 5, ... in the A column. Put the terms of the sequence in the B column. The C column contains the guess for the formula. So if our guess for the $n$th term is $2n^2 - n$ put =2*A1^2−A1 in cell C1 and copy it down the column. The spreadsheet should look like this.

| A | B | C |
|---|---|---|
| 1 | 2 | 1 |
| 2 | 3 | 6 |
| 3 | 8 | 15 |
| 4 | 17 | 28 |
| 5 | 30 | 45 |
| 6 |  | 66 |
| 7 |  | 91 |
| 8 |  | 120 |
| 9 |  | 153 |
| 10 |  | 190 |
| 11 |  |  |
| 12 |  |  |
| 13 |  |  |
| 14 |  |  |
| 15 |  |  |
| 16 |  |  |
| 17 |  |  |
| 18 |  |  |

Adjust the formula in the C column until it matches the terms of the sequence in the B column.

It will speed things up if you put a general quadratic formula, $an^2 + bn + c$, in the C column. Use cells D1, D2 and D3 for $a$, $b$ and $c$. The formula in C1 is
=D$1*A1^2+D$2*A1+D$3
Now the formula can be changed instantly by changing the coefficients in D1, D2 and D3.

> The $ sign prevents the number changing as it is copied down the column.

# Summary

## SUMMARY

- A **linear sequence** (or **arithmetic sequence**) changes by adding a constant amount, called a **common difference**. If the first term is $a$, and the common difference is $d$, then the $n$th term is $a + (n-1)d$.
- An **exponential sequence** (or **geometric sequence**) changes by multiplying by a constant amount, called a **common ratio**. If the first term is $a$, and the common ratio is $r$, then the $n$th term is $ar^{n-1}$.
- If the differences of a sequence form a linear sequence, then the $n$th term has a quadratic formula and the sequence is a **quadratic sequence**.

## Exercise 10A

1. Find the $n$th term of the sequence $5, 7, 9, 11, \ldots$
2. Find the number of dots in the $n$th pattern of the sequence below.

3. The rent of a flat is £4000 per year. It increases by £300 each year. How much is the rent for the $n$th year?
4. Write down the next two terms in the sequence $8, 12, 18, 27, \ldots$.
5. Find the $n$th term of the sequence of question 4.
6. £4000 is invested at 6% compound interest. How much is there after $n$ years?
7. Find the next two terms of the sequence $3, 7, 13, 21, \ldots$.
8. Find the $n$th term of the sequence of question 7.
9. Draw the next pattern in the sequence below.

10. Find the number of dots in the $n$th term of the sequence of question 9.

## Exercise 10B

1. Find the $n$th term of the sequence $12, 11.5, 11, 10.5, \ldots$.
2. Find the number of lines in the $n$th pattern of the sequence below.

3. The area of rain forest in a country is 12 000 000 hectares. It is being cut down at 20 000 hectares each year. How much is there left after $n$ years?
4. Find the next two terms of the sequence $9, 6, 4, 2\frac{2}{3}, \ldots$.
5. Find the $n$th term of the sequence of question 4.
6. The value of a car is decreasing at 25% each year. If it cost £10 000 new, how much is it worth after $n$ years?
7. Find the next two terms of the sequence $19, 16, 11, 4, \ldots$.
8. Find the $n$th term of the sequence of question 7.
9. Draw the next term of the sequence below.

10. Find the number of dots in the $n$th term of the sequence of question 9.

## Exercise 10C

Fibonacci (1170–1250) wrote about this sequence in his book *Liber abbaci*. It came out of the following problem:

A certain man put a pair of rabbits in a place surrounded on all sides by a wall. How many pairs of rabbits can be produced from that pair in a year if it is supposed that every month each pair begets a new pair which from the second month on becomes productive?

In the sequences of this chapter each value was defined in terms of its predecessor. More complicated sequences can be defined in terms of more than one predecessor. In particular, the Fibonacci numbers are defined as follows.

First term = second term = 1. Then each term is the sum of its two predecessors.

So the next three terms are $1 + 1 = 2$, $1 + 2 = 3$, $2 + 3 = 5$.

1 Write out the first ten terms of the sequence. A computer spreadsheet can do this for you.
2 Find the ratio between successive terms i.e. $\dfrac{(n+1)\text{th term}}{n\text{th term}}$. What happens to these ratios as $n$ increases?
3 What happens if you change the formula above? There are several things you could try. Here are some suggestions.
  a Change the initial values of the first two terms.
  b Make each term the sum of its *three* predecessors.

## Exercise 10D

The following sequence is of the first 10 prime numbers.

2  3  5  7  11  13  17  19  23  29

Prime numbers are very important throughout mathematics. For many years mathematicians have tried to find a formula which gives the sequence of prime numbers.

1 One suggestion for a formula to generate primes was $n^2 - n + 41$. Work out this expression for the first few values of $n$. Are they all prime?
2 Are all the values of $n^2 - n + 41$ prime? (Hint: try 41 itself.)
3 Other formulae which produce many primes are below. Test them for the first few values of $n$. Do they always give primes?
  a $n^2 - n + 17$            b $2n^2 + 29$
4 In 1640 Pierre de Fermat (1601–1665) thought that, if $n$ was prime, $2^{2^n} + 1$ would also be prime. Show that this works for $n$ equal to 0, 1, 2 and 3. It also works for $n = 4$. But does it work for $n = 5$? Try dividing by 641.

## Exercise 10E

Here is a simple sequence which has very surprising properties. To investigate it you need the help of a computer. In particular, a spreadsheet can do all the number crunching.

Consider the sequence with first term 0.5, and for which each term is $k$ times its predecessor, times 1 minus its predecessor. So

> first term = 0.5, $(n+1)$th term = $k \times n$th term $\times (1 - n$th term$)$
> (this is called the logistic sequence)

So for example, if $k = 1.5$, the next two terms are

> second term = $1.5 \times 0.5(1 - 0.5) = 0.375$
> third term = $1.5 \times 0.375(1 - 0.375) = 0.3515625$

Enter 0.5 into cell B1. In cell B2 enter =1.5*B1*(1-B1). Copy this down the B column. What happens to the sequence?

Amend the sequence by changing the value of $k$. It will save time if you have $k$ in a separate cell. Put 1.5 in cell A1. Amend B2 to =A$1*B1*(1-B1), and copy down the B column. Now whenever you alter the value of $k$ in A1 the whole B column alters.

Some values of $k$ to try are 2.5, 3, 3.6, 3.83, 3.9.

**You will need:**
- computer with spreadsheet package installed

# 11 Sampling

In the statistical examples so far, you have been presented with the data. How were they obtained? Do they represent fairly the quantity we are trying to measure? For example, the BBC and television companies regularly publish the viewing figures for their programmes. In this chapter we examine the methods for obtaining data. When it is reported that the audience for *Coronation Street* on Wednesday evening was 8 million, how was that figure obtained? Not all 8 million viewers were asked, surely! Did you watch it, and were you asked?

> TV viewing figures are provided by BARB, the Broadcasters' Audience Research Board.

In some cases probability or proportion can be found theoretically. If a fair coin is spun, the probability that it will give Heads is $\frac{1}{2}$. But, if the coin is distorted, then we cannot find the probability by theory. We have to rely on experiment – by spinning the coin many times and finding the proportion of Heads. The proportion is sometimes called the **relative frequency**, i.e. the number of Heads divided by the number of spins.

The more times the experiment is repeated, the more likely it is that the experimental probability will be close to the true proportion. Nothing much can be deduced from an opinion poll of 10 people, and the results from a poll of 100 people are unreliable, but an opinion poll of 1000 people will give good results. Typically, if 40% of the 1000 people support a particular party, then we can be fairly sure (95% sure) that the national proportion supporting the party lies within the range 37% to 43%.

## Exercise 11.1

Mainly revision

1  Harry spun a misshapen coin 20 times, and obtained 12 Heads. He passed the coin over to Lucy, who spun it 80 times and obtained 55 heads.
   a  What is Harry's experimental probability that this coin gives Heads?
   b  What is Lucy's experimental probability that this coin gives Heads?
   c  Combining all the results, what is the experimental probability that the coin gives Heads?
   d  Which probability is most likely to be closest to the true value? Do you think the coin is fair? Discuss with a partner.

2  A dice was rolled 10 times and gave 3 sixes. It was rolled for a further 190 times and gave 30 sixes.
   a  From the first 10 rolls, what is the experimental probability that this dice will give a six?
   b  From the next 190 rolls, what is the experimental probability that this dice will give a six?
   c  From all the rolls, what is the experimental probability that this dice will give a six?
   d  Do you think the dice is fair? Discuss with a partner.

3  Two opinion polls were taken on the same day. The first questioned 500 people and found that 170 of them supported the Purple Party. The second questioned 1500 people and found that 540 of them supported the Purple Party. What is the percentage support for the Purple Party according to
   a  the first poll
   b  the second poll
   c  both polls combined?
   d  Which is most likely to be closer to the support for the party throughout the country?

You can simulate the finding of experimental probability using a calculator or computer.

**Calculator.** On most scientific calculators there is a button which gives a random number between 0 and 1. It may be labelled RND. Suppose a misshapen coin has probability 0.4 of giving Heads. Press the RND button, and count it as a Head if the number is less than 0.4, and a Tail otherwise. Press the button many times, and record the number of Heads you obtained. Find the proportion of Heads you obtained after 10 goes, 20 goes, 50 goes and 100 goes. You will probably find that the proportion gets closer to 0.4.

**Computer.** A spreadsheet function which returns a random number between 0 and 1 is =RAND().
In the A column, fill in the numbers 1 to 100.
In cell B1, enter =IF(RAND()<.4,1,0). This will give 1 (Heads) if the random number is less than 0.4, and 0 (Tails) otherwise. Copy this down to B100.
Use the C column to record the total of Heads, and the D column for the proportion of Heads. So in C1 enter =B1, and in C2 enter =C1+B2. Copy this down to C100.

In D1 enter =C1/A1, and copy down to D100.

You should find that the numbers in the D column get closer to 0.4.

**Note.** This is a probability experiment, and we can never be certain that the results will be as predicted. In 100 goes, the proportion of Heads *should* get closer to 0.4, but it is possible that it does not.

## Samples

> The electorate is all those people qualified to vote.

The set of all the things we are investigating is the **population**. This does not have to be the population of the country – in an opinion poll, for example, the population consists of the electorate. The population does not include those people who are not entitled to vote: those under 18, etc. When we are trying to find a probability or proportion of the population by experiment, the items we investigate, or the people we question, form a **sample**. To find the probability that a coin gives Heads, the sample consists of the spins. To find the support for a particular party, the sample consists of the people in the opinion poll. It is not always easy to select a sample which is representative of what we are trying to investigate.

### Random sample

In a **random sample**, every item has an equal chance of being picked. So in a random opinion poll, every elector has an equal chance of being questioned. You can obtain a random sample by making a list of the items you are investigating, and picking some of them at random. There are several ways of doing this. Here are two.

- Drawing names from a hat. If there are only a few items, they can be written down on pieces of paper which are then drawn from a hat.
- Random numbers. If there are too many items, then it would take too long to write them all down on pieces of paper. Instead pick random numbers, which will then indicate which items are to form the sample.

# 184 SAMPLING

*Example*  Three people are to be chosen from a year group of 80. Show how this could be done using random numbers.

Draw up a list of the 80 students. Pick random numbers between 1 and 99. Pick the students with these numbers, discarding any number greater than 80 or which has already appeared.

Suppose the random numbers are:

    17    pick student number 17
    93    discard
    43    pick student number 43
    17    discard, as already appeared
    66    pick student number 66

So the students numbered 17, 43 and 66 are chosen.

## Exercise 11.2

1  A school contains 900 students. A group of 10 is to be chosen at random to represent student opinion. Show how this can be done using the following random numbers.

014  328  539  943  301  010  328  488
700  836  920  000  573  203  992

2  Use the random number generator on a calculator to find a random sample of 10 from the school of question 1.

3  In the National Lottery, six numbers are chosen at random from the numbers 1 to 49. One Saturday the machine breaks down, and the selection is made by random numbers. If the following random numbers are obtained, what numbers are chosen?

23  00  53  72  06  17  47  83  23  10
73  72  28  82  59  27  38  42  59  60

4  Use the random number generator on a calculator to pick the six winning numbers of the National Lottery. Explain how you made your selection.

### Stratified sample

Sometimes a population is split up into several groups, and we want to ensure that each of the groups is fairly represented in the sample. This might not occur with a random sample, as by chance the sample might contain far too high a proportion of one particular group. A **stratified sample** ensures that the different groups are fairly represented. Within each group, pick the members of the sample at random. Either

- ensure that the numbers in the sample are proportional to the numbers in the population OR
- adjust the results to ensure that each group is represented fairly.

Both methods will be shown in the following examples.

*Examples*  A sports club contains 160 men and 140 women. The management of the club is considering extending the opening hours, and selects a sample of 60 members for their opinion. How many men and women should there be in the sample if both sexes are to be fairly represented?

The ratio of men to women in the club is 160:140, which is the same as 8:7. Divide 60 in the same ratio.

$$60 \times \tfrac{8}{15} = 32 \qquad 60 \times \tfrac{7}{15} = 28$$

There should be 32 men and 28 women in the sample.

---

A sixth form college has 240 arts students and 300 science students. Opinions were canvassed on whether the college should change its name. A sample of 20 arts students was taken, of whom 16 were in favour of the proposal, and a sample of 20 science students was taken, of whom 10 were in favour. Estimate the support for the proposal in the college as a whole.

There are more science students than arts students in the college. As there are equal numbers of arts and science students in the sample, it is not representative of the college as a whole. For each group, use the results for the sample to estimate the number in favour in the whole group. Then divide by the total number of students.

$\tfrac{4}{5}$ of the sample arts students were in favour: $\tfrac{4}{5} \times 240 = 192$
$\tfrac{1}{2}$ of the sample science students were in favour: $\tfrac{1}{2} \times 300 = 150$
$(192 + 150) \div 540 = \tfrac{19}{30}$

The support in the college as a whole is estimated as $\tfrac{19}{30}$.

## Exercise 11.3

1. A library contains 2000 fiction books and 3000 non-fiction books. A sample of 60 is to be taken to look at the condition of the books. What should the sample contain if both types of book are to be fairly represented?
2. A sports club has 300 adult members and 150 junior members. A sample of 30 is to be taken to sound out opinions on the club. What should the sample contain if both age groups are to be fairly represented?
3. An Indian restaurant finds that the ratio between the customers who eat in the restaurant and those who order a 'take-away' is 7:3. The manager wants to ask 40 of the customers about how service could be improved. How could the 40 people be chosen if both groups of customers are to be fairly represented?
4. A video rental shop classifies its stock as follows.

| comedy | drama | foreign | other |
|---|---|---|---|
| 67 | 128 | 59 | 146 |

A sample of 80 is to be taken to find the condition of the videos. What should the sample consist of, to ensure as fair a representation of each group as possible?

5. A factory employs 320 men and 140 women. A proposal to alter the opening hours of the canteen is made. In a sample, out of 20 men, 7 were in favour of the proposal, and out of 25 women, 15 were in favour. Estimate what proportion of the whole workforce are in favour.
6. A vehicle recovery service has 6 000 000 customers, of whom 1 500 000 subscribe to the 'Diamond Spanner' service. The service is considering extending the facilities available, at an increase of the annual subscription. A sample is taken, and 330 out of 500 of the 'Diamond Spanner' subscribers are in favour, while 210 out of 500 of the other customers are in favour. Estimate support for the proposal among the customers as a whole.

### Bias

In practice it is often difficult to select a sample in a truly random fashion. Then there is a danger that one section of the population will be over-represented in the sample, making its results suspect. In this case the sample is **biased**.

*Example*  A change in the road tax system is proposed. A sample of motorists stopping at a petrol station is questioned for their opinion on the change. Explain why the sample is biased.

Though all the people questioned are motorists, those who stop more frequently at petrol stations are more likely to be asked.

The sample is biased in favour of motorists who have a high mileage.

## Exercise 11.4

Franklin D. Roosevelt (1882–1945) became president in 1933, was re-elected for an unprecedented three more terms, and died in office just before the end of the Second World War. He won the 1936 election handsomely, capturing 60.8% of the votes as against Landon's 36.5%.

1  The 1936 US presidential election was between Franklin D. Roosevelt (Democrat) and Alfred Landon (Republican). A poll conducted on the telephone revealed a clear majority in favour of Landon. Explain why the sample was biased.
2  A new bypass round a town is proposed. The following methods of finding local opinion are suggested. Which will result in a biased sample? Give reasons.
   a  Door-to-door enquiries at 11 a.m. on Monday
   b  Door-to-door enquiries at 7 p.m. on Monday
   c  Questioning people as they arrive at the railway station at 6.30 p.m.
   d  Questioning shoppers in the High Street at 11 a.m. on Saturday
3  La Cucuracha, a chain of Mexican fast-food restaurants, is planning to open a branch in Poshton High Street. Before granting permission, the local council seeks out local opinion. Which of the following would result in a biased sample? Give reasons.
   a  Questioning people in the High Street at 10 a.m. on Saturday
   b  Questioning people in the High Street at 10 p.m. on Saturday
   c  Posting a questionnaire to 100 randomly selected households, and considering the replies
   d  Visiting 100 randomly selected households, and questioning the person who answers the door

## Time series

A **time series** is a record of a quantity which is changing over time. The quantity could be financial, such as sales, average income or prices. It could be physical, such as the temperature or the rainfall. Some specific examples of time series are

- the sales, every week, of a Sunday newspaper
- the level of the euro against the dollar, measured monthly
- the sales of new cars, measured quarterly.

To see how the quantity is changing, draw a graph of the quantity against time. Any general behaviour that you can spot from the graph is the **trend**.

*Example* A car dealership keeps records of its quarterly sales of new cars. The results for 12 successive quarters are below. Plot the graph of the sales. What is the trend?

| quarter | 1 | 2 | 3 | 4 | 5 | 6 | 7 | 8 | 9 | 10 | 11 | 12 |
|---------|---|---|---|---|---|---|---|---|---|----|----|----|
| sales   | 10 | 15 | 37 | 8 | 17 | 15 | 42 | 11 | 21 | 19 | 45 | 10 |

The graph is as shown.

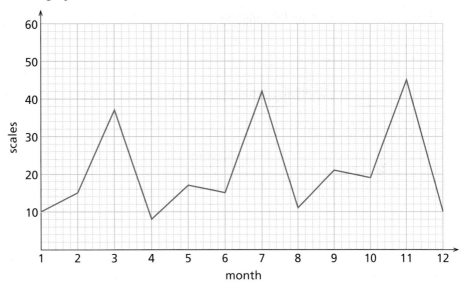

The graph goes up and down, but on the whole the trend is upwards.

## Exercise 11.5

**You will need:**
- graph paper
- the graphs you plot now for exercise 11.6

**1** At its beginning in January 1999, the value of the euro (€) was 1.17 US dollars. The table below gives the value of the euro against the dollar for 24 successive months at two-monthly intervals. Plot the graph of this time series. What is the trend?

| month | 0 | 2 | 4 | 6 | 8 | 10 | 12 | 14 | 16 | 18 | 20 | 22 |
|-------|---|---|---|---|---|----|----|----|----|----|----|----|
| value of € in $ | 1.17 | 1.09 | 1.06 | 1.03 | 1.06 | 1.05 | 1.03 | 0.97 | 0.91 | 0.96 | 0.89 | 0.85 |

**2** The table below gives the rainfall measured every quarter. Plot the graph of this time series. Can you spot a trend?

| quarter | 1 | 2 | 3 | 4 | 5 | 6 | 7 | 8 | 9 | 10 | 11 | 12 |
|---|---|---|---|---|---|---|---|---|---|---|---|---|
| rainfall (mm) | 230 | 100 | 50 | 180 | 270 | 80 | 20 | 230 | 250 | 100 | 40 | 190 |

**3** A shop keeps a record of its sales of ice cream over 12 successive quarters. The results are below. Plot a graph of this time series. Can you spot a trend?

| quarter | 1 | 2 | 3 | 4 | 5 | 6 | 7 | 8 | 9 | 10 | 11 | 12 |
|---|---|---|---|---|---|---|---|---|---|---|---|---|
| sales (£100s) | 3.4 | 7.9 | 17.3 | 3.6 | 2.8 | 8.3 | 20.4 | 3.0 | 3.5 | 9.1 | 19.8 | 3.1 |

**4** The temperature was taken every 8 hours for a four-day period. Plot a graph of this time series. Can you spot a trend?

| hour | 0 | 8 | 16 | 24 | 32 | 40 | 48 | 56 | 64 | 72 | 80 | 88 |
|---|---|---|---|---|---|---|---|---|---|---|---|---|
| temperature (°C) | 8 | 6 | 21 | 9 | 8 | 23 | 8 | 5 | 19 | 10 | 9 | 24 |

## Moving averages

The graph of a time series may be so jagged that it is difficult to draw any useful conclusions from it, i.e. it is difficult to spot any trend. There could be two reasons.

- The numbers could vary randomly.
- There might be a seasonal variation, i.e. the numbers might be greater in one period of the year than another. For example, sales of ice cream are higher in the summer than in the winter.

We can smooth out the jaggedness by taking the average of several successive figures. This is called a **moving average**. The graph of the moving average will smooth out the random fluctuations and, if chosen correctly, will eliminate the seasonal variation. The number of figures chosen for the moving average is the number of **points**. A 4-point moving average takes the mean of four successive numbers.

When plotting a graph of the moving average, plot each value at the middle of its time period. If we have a moving average of 125 for the third, fourth and fifth week, plot at 125 on the vertical axis and 4 on the horizontal axis.

*Example* For the car sales example above, find the 4-point moving averages. Plot these moving averages on the same graph. What can you say about the trend?

By taking 4-point moving averages, we are finding the average for the whole year. This eliminates the distortion due to the very high sales in the third quarter of each year.

The moving averages are

$\frac{1}{4}(10 + 15 + 37 + 8) = 17.5$
$\frac{1}{4}(15 + 37 + 8 + 17) = 19.25$
$\frac{1}{4}(37 + 8 + 17 + 15) = 19.25$
$\frac{1}{4}(8 + 17 + 15 + 42) = 20.5$
$\frac{1}{4}(17 + 15 + 42 + 11) = 21.25$
$\frac{1}{4}(15 + 42 + 11 + 21) = 22.25$
$\frac{1}{4}(42 + 11 + 21 + 19) = 23.25$
$\frac{1}{4}(11 + 21 + 19 + 45) = 24$
$\frac{1}{4}(21 + 19 + 45 + 10) = 23.75$

So the moving average for the first four quarters is 17.5. The middle of the time interval is at 2.5. So plot at (2.5, 17.5). Plot for all the other moving averages.

The graph of the moving averages is shown on the original graph of the time series. The upwards trend is much more evident.

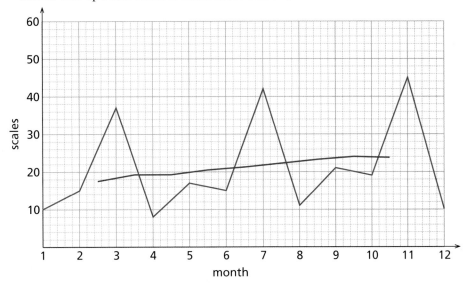

# Exercise 11.6

**You will need:**
- your graphs from exercise 11.5

Questions 1–4 refer to the situations in questions 1–4 of exercise 11.5.

1 Find 3-point moving averages for the value of the euro against the dollar. Plot these moving averages on the same graph as for the original values. Can you spot a trend?
2 Find 4-point moving averages for the rainfall. Plot these moving averages on the same graph as for the original values. Can you spot a trend?
3 Find 4-point moving averages for the sales of ice cream. Plot these moving averages on the same graph as for the original values. Can you spot a trend?
4 Find 3-point moving averages for the temperature. Plot these moving averages on the same graph as for the original values. Can you spot a trend?
5 A newspaper is published from Monday to Saturday inclusive. Moving averages of the daily sales figures are calculated. What should the number of points be for these moving averages?

# SAMPLING

## SUMMARY

- If a probability cannot be found from theory, then it may be possible to find it by experiment. The more the experiment is repeated, the more reliable its results should be.
- Suppose we want to find out a fact about a large **population**. We investigate a smaller part of the population, called a **sample**.
- In a **random sample**, every member of the population has an equal chance of being selected.
- Suppose the population is divided into different groups. A **stratified sample** contains groups for each of the population groups. It is likely to lead to more reliable results.
- If a sample is not representative of the population, it is **biased**.
- A **time series** is a quantity changing over time. The graph of a time series may show a **trend**.
- A **moving average** of a time series is the average taken over a fixed period. The graph of the moving average smoothes out the random fluctuations and seasonal variation of the original time series. The number of figures chosen for the moving average is the number of **points**.

## Exercise 11A

1. Paul drops a drawing pin ten times, and it lands point upwards four times. Alison drops the same pin 70 times, and it lands point upwards 17 times. Find the experimental probability that the pin lands point upwards
   a from Paul's results
   b from Alison's results
   c from all the results.
   Which is most likely to be closest to the true probability?

2. An internet service provider has customers which it numbers 1 to 4 434 000, in order of joining the service. It wishes to select randomly a panel of 1000 customers to be consulted on improvements to the service. Explain how this could be done using random numbers.

3. A college contains 240 male and 360 female students. A group of 30 will be selected to represent student opinion. How should they be selected if both sexes are to be fairly represented?

4. A music shop classifies its stock of CDs as follows.

   | pop | classical | other |
   |---|---|---|
   | 850 | 560 | 720 |

   A sample was taken to investigate the condition of the CD cases. Of ten pop CDs, two had cases in bad condition; of ten classical CDs the number was six; and of ten 'other' the number was three. Estimate the proportion of CDs with cases in bad condition.

5. Refer to question 2. Explain why a sample containing the customers numbered 1 to 1000 would be biased.

6. The question 'Did you enjoy your school days?' is put to a group of university students. Is this an unbiased sample of school leavers?

7. A bookshop records its sales for 12 successive quarters. The results are below. Plot a graph for this time series.

   You will need:
   - graph paper

   | quarter | 1 | 2 | 3 | 4 | 5 | 6 | 7 | 8 | 9 | 10 | 11 | 12 |
   |---|---|---|---|---|---|---|---|---|---|---|---|---|
   | sales (1000s) | 8.1 | 10.3 | 9.2 | 23.8 | 7.3 | 9.2 | 9.0 | 22.9 | 7.1 | 8.2 | 8.8 | 20.1 |

8. For the data in question 7, find 4-point moving averages.
9. Plot the moving averages you found for question 8 on your graph for question 7.
10. Describe any trend in book sales.

## Exercise 11B

1. Chris asked 20 people who they would vote for at the next election, and 7 said they would support the Reactionary Party. Neela asked 80 people, and found 51 supporters of the Reactionary Party. Estimate the support for the Reactionary Party
   a from Chris' results
   b from Neela's results
   c from the combined results.
   Which is most likely to be closest to the true level of support?

2. There are 87 members of a club, numbered 1 to 87. Four are to be chosen at random to form a committee. Make the selection using the following random numbers.

   03  82  91  00  82  48  57

3. A computer firm keeps a database of its customers. It has 320 000 who use the computer at work, and 180 000 who use it at home. A group of 200 is to be selected to give feedback on improvements to the computer. How should the group be chosen to ensure that both sets of users are fairly represented?

4. The adult inhabitants of a town are classified under three age ranges as follows.

   | under 30 | 30 to 50 | over 50 |
   |---|---|---|
   | 24 000 | 42 000 | 34 000 |

   A sample was taken to find the extent of unemployment. The results are

   |  | under 30 | 30 to 50 | over 50 |
   |---|---|---|---|
   | employed | 23 | 31 | 12 |
   | unemployed | 7 | 9 | 28 |

   Estimate the proportion of unemployed in the town as a whole.

5. A questionnaire about the importance of computers in everyday life was sent by e-mail to 1000 people. Explain why this would be a biased sample.

6. A town council sent out 20 000 leaflets asking whether the services it provided were satisfactory. Of the 4000 responses, 2800 said that the services were not satisfactory. Are the people who responded an unbiased sample of the people who received the leaflet?

7. The table below gives the amount of snowfall in 10 successive years. Plot a graph for this time series.

   **You will need:**
   - graph paper

   | year | 1 | 2 | 3 | 4 | 5 | 6 | 7 | 8 | 9 | 10 |
   |---|---|---|---|---|---|---|---|---|---|---|
   | snowfall (cm) | 0 | 83 | 47 | 33 | 0 | 58 | 21 | 8 | 41 | 30 |

8. For the data in question 7, find 3-point moving averages.
9. Plot the moving averages you found for question 8 on your graph for question 7.
10. Describe any trend in snowfall.

## Exercise 11C

How many students are there in your school or college? If it is large, then it would be difficult to canvass everyone's opinion on a topic that interests you. Obtain a school list, and from that obtain a random sample of size 20. Question the people in the sample about the topic.

## Exercise 11D

In the section on time series, all the data were given to you. Collect data of your own about a quantity that varies over time. Plot these data, and also plot the moving averages of a suitable order. Can you find a trend?

**You will need:**
- graph paper

## Exercise 11E

There is a lot of number crunching in finding moving averages. A spreadsheet can do all the arithmetic for you, and all the graph plotting for you.

Put the data of the time series from exercise 11D in the A column. The B column will hold the moving averages. If they are 4-point moving averages, in B4 enter =(A1+A2+A3+A4)/4 and copy down the B column. Plot the graphs of the original data and of the moving averages.

**You will need:**
- computer with spreadsheet package installed

# 12 Vectors

Recall that a **vector** is a quantity that has direction as well as **magnitude** (size or length). A scalar is a quantity that has magnitude but no direction. Here are some quantities, described in terms of whether or not they are vectors.

- Force: a vector, as we want to know the direction a force is acting in
- Velocity: a vector, as it gives the direction a body is moving in, as well as its speed
- Electrical current: a vector, as we want to know the direction a current is flowing in
- Temperature: a scalar, as it makes no sense to say: '15° Celsius acting upwards'
- Money: a scalar, as it makes no sense to say: 'I earned £250 in a northerly direction'

So in all the examples and use of vectors that we have, it is important to know the direction as well as the magnitude.

## Vectors as translations

In a translation, points are moved in a particular direction by a particular distance. Hence a translation can be represented by a vector, as it involves a direction as well as a magnitude. If the translation moves points a distance $x$ in the horizontal direction, and $y$ in the vertical direction, then it is written as $\begin{pmatrix} x \\ y \end{pmatrix}$.

*Example* Find the vector for the translation which takes (3, 4) to (6, 1). Find where this translation takes (−3, 7).

The translation moves 3 to the right and 3 down.
The vector is $\begin{pmatrix} 3 \\ -3 \end{pmatrix}$.
Add 3 to −3, and subtract 3 from 7.
The translation takes (−3, 7) to (0, 4).

**Remember:**
Negative numbers mean movement left or downwards.
Positive numbers mean movement right or upwards.

## Exercise 12.1

Mainly revision

1  A translation takes (1, 7) to (5, 2). Write down the vector for this translation.
2  A translation takes (4, −3) to (−1, 8). Write down the vector for this translation.
3  A translation has vector $\begin{pmatrix} 1 \\ -4 \end{pmatrix}$. Where does the translation take (2, 3)?
4  A translation has vector $\begin{pmatrix} 3 \\ 8 \end{pmatrix}$. Where does the translation take (−1, −3)?

# Arithmetic and geometry of vectors

Vectors can be added or subtracted, or multiplied by scalars. These operations can be done either arithmetically or geometrically.

> There is more than one way of writing a vector. When typing or printing, it is usual to write vectors in bold, as **a**. When writing, it is easier to write them underlined, as a̲. Use whichever notation you find easier and clearer.

## Adding

**Arithmetically.** To add two vectors, add the corresponding terms.

$$\begin{pmatrix} 1 \\ 2 \end{pmatrix} + \begin{pmatrix} 2 \\ -1 \end{pmatrix} = \begin{pmatrix} 3 \\ 1 \end{pmatrix}$$

**Geometrically.** To add **a** and **b**: draw **a**, then draw **b** from the head of **a**. Then draw the third side of the triangle as shown. This gives **a** + **b**.

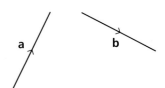

For the vectors $\mathbf{a} = \begin{pmatrix} 1 \\ 2 \end{pmatrix}$ and $\mathbf{b} = \begin{pmatrix} 2 \\ -1 \end{pmatrix}$ above, **a** moves 1 right and 2 up. This is followed by **b**, which moves 2 right and 1 down. The result is **a** + **b**, which moves 3 right and 1 up.

It does not matter which vector comes first, **a** + **b** is the same as **b** + **a**. When adding three or more vectors, it does not matter which order you add them in. With **a** + **b** + **c**, it does not matter whether you add **a** and **b** first or **b** and **c**.

> **a** + **b** is the same as **b** + **a**.
> (**a** + **b**) + **c** is the same as **a** + (**b** + **c**).

## Subtracting

**Arithmetically.** To subtract one vector from another, subtract the corresponding terms.

$$\begin{pmatrix} 1 \\ 2 \end{pmatrix} - \begin{pmatrix} 3 \\ 4 \end{pmatrix} = \begin{pmatrix} -2 \\ -2 \end{pmatrix}$$

**Geometrically.** To subtract **b** from **a**, draw a vector from the head of **b** to the head of **a**. This gives **a** − **b**.

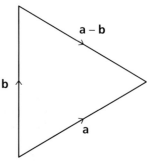

Subtraction can also be thought of as follows: draw **a**, then draw **b** from the head of **a**, but in the opposite direction. Complete the triangle for **a** − **b**.

For the vectors $\mathbf{a} = \begin{pmatrix} 1 \\ 2 \end{pmatrix}$ and $\mathbf{b} = \begin{pmatrix} 3 \\ 4 \end{pmatrix}$ above, **a** moves 1 right and 2 up. To subtract **b**, moves 3 *left* and 4 *down*. The result is **a** − **b**, which moves 2 left and 2 down.

# Arithmetic and geometry of vectors

## Multiplying by a scalar

**Arithmetically.** To multiply a vector by a scalar, multiply both terms.

$$6 \times \begin{pmatrix} 1 \\ 2 \end{pmatrix} = \begin{pmatrix} 6 \\ 12 \end{pmatrix}$$

**Geometrically.** To multiply **a** by a scalar $k$, draw a vector in the same direction as **a** but $k$ times as long. This gives $k\mathbf{a}$.

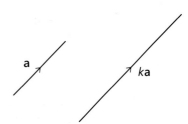

*Examples* The diagram shows vectors **u** and **v**.

Draw
i) $\mathbf{u} + \mathbf{v}$ ii) $2\mathbf{u}$ iii) $2\mathbf{u} - \mathbf{v}$

i) Move **v** so that its tail is on the head of **u**. Then join the third side of the triangle, as shown.

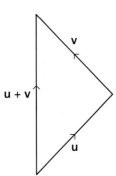

ii) Double the length of **u**, as shown.

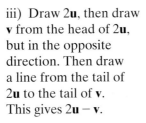

iii) Draw $2\mathbf{u}$, then draw **v** from the head of $2\mathbf{u}$, but in the opposite direction. Then draw a line from the tail of $2\mathbf{u}$ to the tail of **v**. This gives $2\mathbf{u} - \mathbf{v}$.

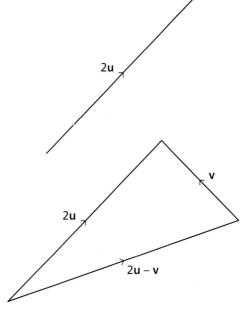

# 196 VECTORS

Let $\mathbf{u} = \begin{pmatrix} 4 \\ -3 \end{pmatrix}$ and $\mathbf{v} = \begin{pmatrix} -1 \\ 7 \end{pmatrix}$. Find $2\mathbf{u} - 3\mathbf{v}$.

Multiply the terms of $\mathbf{u}$ by 2 and the terms of $\mathbf{v}$ by 3, then subtract.

$$2\mathbf{u} - 3\mathbf{v} = 2\begin{pmatrix} 4 \\ -3 \end{pmatrix} - 3\begin{pmatrix} -1 \\ 7 \end{pmatrix} = \begin{pmatrix} 11 \\ -27 \end{pmatrix}$$

With $\mathbf{u}$ and $\mathbf{v}$ as in the previous example, find $x$ and $y$ such that $x\mathbf{u} + y\mathbf{v} = \begin{pmatrix} 5 \\ 15 \end{pmatrix}$.

We have $x\begin{pmatrix} 4 \\ -3 \end{pmatrix} + y\begin{pmatrix} -1 \\ 7 \end{pmatrix} = \begin{pmatrix} 5 \\ 15 \end{pmatrix}$. From the top line and the bottom line

$$4x - y = 5 \qquad [1]$$
$$-3x + 7y = 15 \qquad [2]$$

Multiply equation [1] by 7 and add to equation [2].

$$28x - 7y = 35$$
$$25x = 50$$
$$x = 2$$

Substitute in [1].

$$4 \times 2 - y = 5$$
$$y = 3$$

$x = 2$ and $y = 3$.

# Exercise 12.2

**1** Copy this diagram of $\mathbf{u}$ and $\mathbf{v}$. On your diagram draw the vectors for
   i) $\mathbf{u} + \mathbf{v}$   ii) $\mathbf{u} - \mathbf{v}$   iii) $3\mathbf{v}$

**2** Let $\mathbf{u} = \begin{pmatrix} 3 \\ 2 \end{pmatrix}$ and $\mathbf{v} = \begin{pmatrix} 4 \\ -3 \end{pmatrix}$. Find

   i) $\mathbf{u} + \mathbf{v}$   ii) $2\mathbf{u} - 3\mathbf{v}$

**3** With $\mathbf{u}$ and $\mathbf{v}$ as in question 2, find $x$ and $y$ such that $x\mathbf{u} + y\mathbf{v} = \begin{pmatrix} 32 \\ -7 \end{pmatrix}$.

## Arithmetic and geometry of vectors

### Magnitude

The magnitude of a vector is its size or length. We show the magnitude of a vector by putting lines on either side, as |**a**|. The magnitude is found by Pythagoras' theorem.

$$\text{Magnitude of } \begin{pmatrix} x \\ y \end{pmatrix} = \left| \begin{pmatrix} x \\ y \end{pmatrix} \right|$$
$$= \sqrt{x^2 + y^2}$$

**Remember:**
$c^2 = a^2 + b^2$

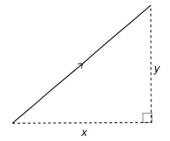

*Examples* Find the magnitude of $\begin{pmatrix} 3 \\ -5 \end{pmatrix}$.

Use the formula, putting $x = 3$ and $y = -5$.

$$\left| \begin{pmatrix} 3 \\ -5 \end{pmatrix} \right| = \sqrt{3^2 + (-5)^2} = \sqrt{34}$$

The magnitude is $\sqrt{34}$.

Let **u** have magnitude 6 units and direction 040°, and let **v** have magnitude 11 units and direction 070°. Find the magnitude and direction of **u** + **v**.

**u** + **v** is the third side of the triangle as shown. The (obtuse) angle between **u** and **v** is 150°. Use the cosine rule to find the magnitude of **u** + **v**.

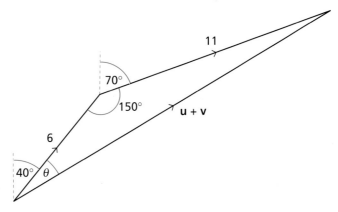

$|\mathbf{u} + \mathbf{v}|^2 = 6^2 + 11^2 - 2 \times 6 \times 11 \times \cos 150° = 271.3$

**Remember:**
$a^2 = b^2 + c^2 - 2bc \cos A$

The magnitude of **u** + **v** is 16.5 units.

For the direction, use either the sine rule or the cosine rule. Let the angle between **u** and **u** + **v** be $\theta$. Using the sine rule

**Remember:**
$\dfrac{a}{\sin A} = \dfrac{b}{\sin B} = \dfrac{c}{\sin C}$

$$\frac{\sin \theta}{11} = \frac{\sin 150°}{16.5}$$

Hence $\theta = 19.5°$. Add on to 40°.
The direction of **u** + **v** is 059.5°.

Check: the magnitude of **u** + **v** is a bit less than 6 + 11, which is correct. Its direction is between those of **u** and **v**, and closer to **v**, which is also correct.

## Exercise 12.3

**1** Find the magnitudes of these vectors.

i) $\begin{pmatrix} 2 \\ 7 \end{pmatrix}$

ii) $\begin{pmatrix} 4 \\ 1 \end{pmatrix}$

iii) $\begin{pmatrix} -2 \\ 6 \end{pmatrix}$

iv) $\begin{pmatrix} 8 \\ -7 \end{pmatrix}$

v) $\begin{pmatrix} -4 \\ -3 \end{pmatrix}$

vi) $\begin{pmatrix} 0.3 \\ 0.7 \end{pmatrix}$

**2** Let **u** have magnitude 4 units and direction 060°, and let **v** have magnitude 5 units and direction 110°. Find the magnitude and direction of
i) **u** + **v**
ii) **u** − **v**
iii) 3**u** + 2**v**
iv) 4**u** − 3**v**

**3** Let **u** have magnitude 7 units and direction 030°, and let **v** have magnitude 8 units and direction 320°. Find $x$, if the direction of $x$**u** + **v** is due North.

**4** With **u** and **v** as in question 3, find $x$ if the magnitude of $x$**u** + **v** is 14 units.

# Vectors and coordinate geometry

Vectors can be used in coordinate geometry problems. In particular, we use the following definitions and results.

If $A$ and $B$ are points, the vector from $A$ to $B$ is written $\overrightarrow{AB}$.

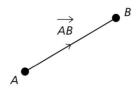

The **position vector** of a point $A$ is the vector which goes from the origin to $A$, i.e. $\overrightarrow{OA}$.

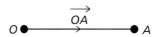

If $A$ is at $(x, y)$, then the position vector of $A$ is $\begin{pmatrix} x \\ y \end{pmatrix}$. So the position vector of $(1, 2)$ is $\begin{pmatrix} 1 \\ 2 \end{pmatrix}$. It is easy to convert between the coordinates of a point and its position vector.

Here are two results about position vectors.

- *If $A$ and $B$ have position vectors $\mathbf{a}$ and $\mathbf{b}$, then*
  *i) $\overrightarrow{AB} = \mathbf{b} - \mathbf{a}$  ii) the midpoint of $AB$ has position vector $\frac{1}{2}(\mathbf{a} + \mathbf{b})$*

**Proof**

i) Suppose we go from $A$ to $B$. Go the long way, via $O$.

$$\overrightarrow{AB} = \overrightarrow{AO} + \overrightarrow{OB} = -\overrightarrow{OA} + \overrightarrow{OB} = -\mathbf{a} + \mathbf{b} = \mathbf{b} - \mathbf{a}$$

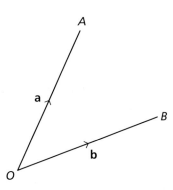

ii) Let $M$ be the midpoint of $AB$. Then the position vector of $M$ is

$$\overrightarrow{OM} = \overrightarrow{OA} + \tfrac{1}{2}\overrightarrow{AB}$$
$$= \mathbf{a} + \tfrac{1}{2}(\mathbf{b} - \mathbf{a})$$
$$= \tfrac{1}{2}(\mathbf{a} + \mathbf{b})$$

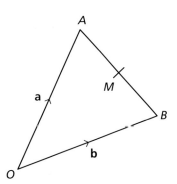

**Notes**

i) Be sure to get this the right way round. In going from $A$ to $B$ you subtract the position vector of $A$ from that of $B$.

ii) The position vector of $M$ is the mean of those of $A$ and $B$.

*Examples* Let $A$, $B$ and $D$ be at $(1, 1)$, $(7, 3)$ and $(2, 11)$ respectively. Find the point $C$ such that $ABCD$ is a parallelogram.

In a parallelogram, opposite sides are equal and parallel. Hence $\vec{BC} = \vec{AD}$.

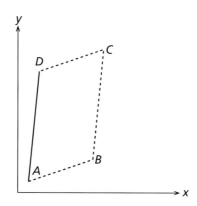

Hence $\vec{BC} = \begin{pmatrix} 2 \\ 11 \end{pmatrix} - \begin{pmatrix} 1 \\ 1 \end{pmatrix} = \begin{pmatrix} 1 \\ 10 \end{pmatrix}$. Add this to the position vector of $B$.

$\begin{pmatrix} 7 \\ 3 \end{pmatrix} + \begin{pmatrix} 1 \\ 10 \end{pmatrix} = \begin{pmatrix} 8 \\ 13 \end{pmatrix}$. This is the position vector of $C$.

$C$ is at $(8, 13)$.

Check: try the other sides. $\vec{AB} = \begin{pmatrix} 7 \\ 3 \end{pmatrix} - \begin{pmatrix} 1 \\ 1 \end{pmatrix} = \begin{pmatrix} 6 \\ 2 \end{pmatrix}$, and $\vec{DC} = \begin{pmatrix} 8 \\ 13 \end{pmatrix} - \begin{pmatrix} 2 \\ 11 \end{pmatrix} = \begin{pmatrix} 6 \\ 2 \end{pmatrix}$.

So the other pair of opposite sides are also equal and parallel.

---

Let $A$, $B$ and $C$ be at $(1, 3)$, $(5, 5)$ and $(7, 6)$ respectively. Show that $A$, $B$ and $C$ lie on a straight line.

$\vec{AB} = \begin{pmatrix} 4 \\ 2 \end{pmatrix}$ and $\vec{BC} = \begin{pmatrix} 2 \\ 1 \end{pmatrix}$. Note that $\vec{AB} = 2\vec{BC}$ and hence they are parallel. As these line segments have $B$ in common, they must form part of a straight line.

$A$, $B$ and $C$ lie on a straight line.

## Exercise 12.4

**1** Let $P$, $Q$ and $R$ be at $(1, 3)$, $(2, -3)$ and $(4, 9)$ respectively. Find the following vectors.
 i) $\vec{PQ}$
 ii) $\vec{PR}$
 iii) $\vec{RQ}$

**2** The position vector of $A$ is $\begin{pmatrix} 3 \\ 5 \end{pmatrix}$, and $\vec{AB} = \begin{pmatrix} -7 \\ 3 \end{pmatrix}$. Find the position vector of $B$.

**3** The position vector of $C$ is $\begin{pmatrix} -1 \\ 8 \end{pmatrix}$, and $\vec{DC} = \begin{pmatrix} 11 \\ 23 \end{pmatrix}$. Find the position vector of $D$.

**4** Show that the four points $A(2, 5)$, $B(6, 10)$, $C(15, 13)$ and $D(11, 8)$ form a parallelogram.

5  By considering the lengths of the sides of $ABCD$ of question 4, find out whether it is a rhombus.
6  By considering the lengths of the diagonals of $ABCD$ of question 4, find out whether it is a rectangle.

> The diagonals of a rectangle are equal in length.

7  Show that the points $P(2, 3)$, $Q(6, 9)$, $R(9, 7)$ and $S(5, 1)$ form a rectangle. Do they form a square?
8  Three points are $A(2, 3)$, $B(7, 4)$ and $C(4, 11)$. Find
   i) point $D$, such that $ABDC$ is a parallelogram
   ii) point $E$, such that $AECB$ is a parallelogram.
9  Show that the points $(3, 5)$, $(4, 8)$ and $(7, 17)$ lie on a straight line.
10 Find $k$, given that $(1, 5)$, $(3, 11)$ and $(k, 14)$ lie on a straight line.
11 Find $h$, given that $(1, 6)$, $(h, h)$ and $(9, 19)$ lie on a straight line.

# Vectors in pure geometry

We round off this chapter by using vectors in pure geometry. The results are not dependent on any coordinate system. First recall the following.

- If $A$ is a point, then the position vector of $A$ is the vector from the origin to $A$, i.e. it is $\overrightarrow{OA}$.
- Two vectors are parallel if one is a scalar multiple of the other.
- Suppose points $A$ and $B$ have position vectors $\mathbf{a}$ and $\mathbf{b}$. Then the position vector of the midpoint of $AB$ is the mean of $\mathbf{a}$ and $\mathbf{b}$, i.e. it is $\tfrac{1}{2}(\mathbf{a} + \mathbf{b})$.

*Example*  Let $A$, $B$, $C$ and $D$ be any four points. Let the midpoints of $AB$, $BC$, $CD$ and $DA$ be $P$, $Q$, $R$ and $S$ respectively. Then $PQRS$ is a parallelogram.

Let the position vectors of $A$, $B$, $C$ and $D$ be $\mathbf{a}$, $\mathbf{b}$, $\mathbf{c}$, and $\mathbf{d}$ respectively. Then the position vectors of $P$, $Q$, $R$ and $S$ are

$$\tfrac{1}{2}(\mathbf{a} + \mathbf{b}), \tfrac{1}{2}(\mathbf{b} + \mathbf{c}), \tfrac{1}{2}(\mathbf{c} + \mathbf{d}) \text{ and } \tfrac{1}{2}(\mathbf{d} + \mathbf{a}) \text{ respectively}$$

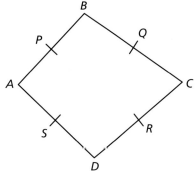

The vector $\overrightarrow{PQ}$ is $\tfrac{1}{2}(\mathbf{b} + \mathbf{c}) - \tfrac{1}{2}(\mathbf{a} + \mathbf{b}) = \tfrac{1}{2}(\mathbf{c} - \mathbf{a})$.
The vector $\overrightarrow{SR}$ is $\tfrac{1}{2}(\mathbf{c} + \mathbf{d}) - \tfrac{1}{2}(\mathbf{d} + \mathbf{a}) = \tfrac{1}{2}(\mathbf{c} - \mathbf{a})$.
Hence $\overrightarrow{PQ} = \overrightarrow{SR}$. Similarly $\overrightarrow{SP} = \overrightarrow{RQ}$. So the opposite sides of $PQRS$ are equal and parallel.

$PQRS$ is a parallelogram.

**Note.** This is a rather stunning result. It applies even when the original four points are not in the same plane. They could be any four points in three-dimensional space.

## Exercise 12.5

1. Let $OAB$ be any triangle. Let $\overrightarrow{OA} = \mathbf{a}$ and $\overrightarrow{OB} = \mathbf{b}$. Let $M$ and $N$ be the midpoints of $OA$ and $OB$ respectively.
   i) Find $\overrightarrow{OM}$ and $\overrightarrow{ON}$.
   ii) Find $\overrightarrow{MN}$ and $\overrightarrow{AB}$.
   iii) What is the relationship between $MN$ and $AB$?

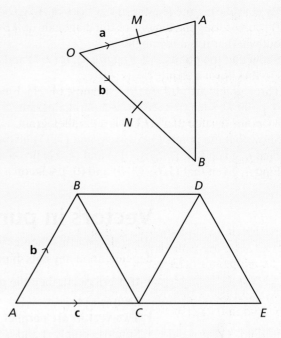

2. The diagram shows three equilateral triangles arranged in a row. $\overrightarrow{AB} = \mathbf{b}$ and $\overrightarrow{AC} = \mathbf{c}$. Express each of the following in terms of $\mathbf{b}$ and $\mathbf{c}$.
   i) $\overrightarrow{BC}$
   ii) $\overrightarrow{AE}$
   iii) $\overrightarrow{BD}$
   iv) $\overrightarrow{DE}$

3. $ABCD$ is a parallelogram, with $\overrightarrow{AB} = \mathbf{b}$ and $\overrightarrow{AD} = \mathbf{d}$. $X$ and $Y$ are the midpoints of $AB$ and $CD$ respectively. Find in terms of $\mathbf{b}$ and $\mathbf{d}$:
   i) $\overrightarrow{AX}$
   ii) $\overrightarrow{AY}$
   iii) $\overrightarrow{XY}$
   What can you say about the line $XY$?

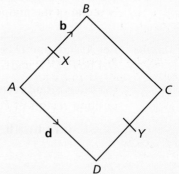

4. $OABC$ is a quadrilateral in which the position vectors of $A$, $B$ and $C$ are $\mathbf{a}$, $\mathbf{b}$ and $\mathbf{c}$ respectively. Let $X$ and $Y$ be the midpoints of $OB$ and $AC$ respectively.
   i) Write $\overrightarrow{OX}$ in terms of $\mathbf{b}$.
   ii) Write $\overrightarrow{OY}$ in terms of $\mathbf{a}$ and $\mathbf{c}$.

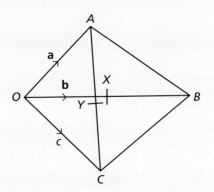

Suppose now that $OABC$ is a parallelogram.
iii) Write $\mathbf{b}$ in terms of $\mathbf{a}$ and $\mathbf{c}$.
iv) Use your answers to i) and iii) to write $\overrightarrow{OX}$ in terms of $\mathbf{a}$ and $\mathbf{c}$.
What can you say about the points $X$ and $Y$?

5 Show that the points with position vectors 2**a**, **a** + **b** and 2**b** are in a straight line.
6 Four points have position vectors **a**, **b**, **a** + **b** and **a** − **b**. Which three of these points are in a straight line?
7 In $\triangle ABC$, $\overrightarrow{AB}=\mathbf{b}$ and $\overrightarrow{AC}=\mathbf{c}$. $X$ lies on $AB$ with $AX = \tfrac{3}{4}AB$. $Y$ and $Z$ are the midpoints of $AC$ and $CX$ respectively.

i) Find in terms of **b** and **c**: $\overrightarrow{AX}, \overrightarrow{AY}, \overrightarrow{AZ}$ and $\overrightarrow{ZY}$.

ii) What can you say about $ZY$?

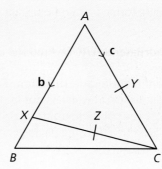

8 In triangle $OAB$, the position vectors of $A$ and $B$ are **a** and **b** respectively. The midpoint of $AB$ is $X$. $Y$ lies on $OA$ with $OY = \tfrac{2}{3}OA$. Extend $OB$ to $Z$ so that $BZ = OB$.

i) Find the position vectors of $X$, $Y$ and $Z$.

ii) Find $\overrightarrow{XY}$ and $\overrightarrow{YZ}$.

iii) What can you say about the three points $X$, $Y$ and $Z$?

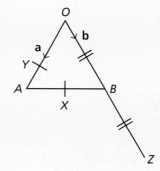

9 $OABCDE$ is a regular hexagon, with the position vectors of $A$ and $E$ being **a** and **e** respectively. Find each of the following in terms of **a** and **e**.

$\overrightarrow{BC} \quad \overrightarrow{CD} \quad \overrightarrow{EB} \quad \overrightarrow{OB}$

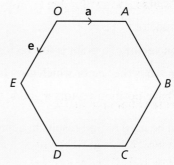

10 Let the position vectors of $A$, $B$ and $C$ be **a**, **b** and **c** respectively. Let $L$, $M$ and $N$ be the midpoints of $BC$, $CA$ and $AB$ respectively. The lines $AL$, $BM$ and $CN$ are called the **medians** of $\triangle ABC$.

i) Find the position vector of the point $G_1$ on $AL$ which is $\tfrac{2}{3}$ of the way from $A$ to $L$. (i.e. $OG_1 = OA + \tfrac{2}{3}AL$)

ii) Find the position vector of the point $G_2$ on $BM$ which is $\tfrac{2}{3}$ of the way from $B$ to $M$.

iii) Find the position vector of the point $G_3$ on $CN$ which is $\tfrac{2}{3}$ of the way from $C$ to $N$.

iv) What can you conclude?

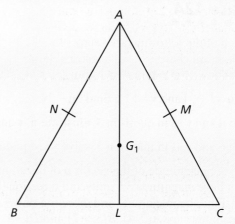

## Exercise 12.6

You will need:
- four sticks, all different lengths
- thread

Look at the example immediately before exercise 12.5, in which we proved that four midpoints formed a parallelogram. Here you verify the result practically.

1. Draw any quadrilateral $ABCD$. Find the midpoints $P, Q, R$ and $S$ of the four sides. Show that $PQRS$ is a parallelogram.
2. This is a three-dimensional verification. Take four sticks of unequal length, fix them together so that they form a quadrilateral which is not in one plane. Join the midpoints of the four sticks with thread, and show that it is a parallelogram.
3. If you have access to a drawing package, it can also be used to verify the result.

## SUMMARY

- A **vector** is a quantity with direction as well as **magnitude** (size or length).
- Vectors can be added, subtracted or multiplied by a constant, either by combining their components or geometrically.
- The magnitude of $\begin{pmatrix} x \\ y \end{pmatrix}$ is $\sqrt{x^2 + y^2}$.
- The **position vector** of a point is the vector which goes from the origin to the point. If $A$ and $B$ are points, then $\overrightarrow{AB}$ is the vector which goes from $A$ to $B$.
- If $A$ and $B$ have position vectors $\mathbf{a}$ and $\mathbf{b}$, then $\overrightarrow{AB} = \mathbf{b} - \mathbf{a}$ and the midpoint of $AB$ has position vector $\frac{1}{2}(\mathbf{a} + \mathbf{b})$.
- Vectors can be used to find results in geometry, whether with or without coordinates.

## Exercise 12A

1. Which of the following are vectors?
   i) electrical charge   ii) electrical field
2. Write down the vector which translates $(4, 7)$ to $(1, 11)$.
3. Let $\mathbf{u} = \begin{pmatrix} 3 \\ -1 \end{pmatrix}$ and $\mathbf{v} = \begin{pmatrix} 2 \\ 3 \end{pmatrix}$. Find $3\mathbf{u} - 2\mathbf{v}$.
4. With $\mathbf{u}$ and $\mathbf{v}$ as in question 3, illustrate $\mathbf{u}, \mathbf{v}$ and $\mathbf{u} + \mathbf{v}$.
5. With $\mathbf{u}$ and $\mathbf{v}$ as in question 3, solve the equation $x\mathbf{u} + y\mathbf{v} = \begin{pmatrix} 13 \\ -8 \end{pmatrix}$.
6. Find the magnitude of the vector $\mathbf{u}$ of question 3.
7. Let $\mathbf{u}$ have magnitude 5.5 units and direction $080°$, and let $\mathbf{v}$ have magnitude 6.5 units and direction $160°$. Find the magnitude and direction of $\mathbf{u} + \mathbf{v}$.
8. Three points are $A(2, 4), B(3, -2)$ and $D(-4, -5)$. Find $C$, given that $ABCD$ is a parallelogram.
9. Is $ABCD$ of question 8
   i) a rhombus   ii) a rectangle?

10 $ABC$ is a triangle, with $\vec{AB} = \mathbf{b}$ and $\vec{AC} = \mathbf{c}$. $X$ and $Y$ are points on $AB$ and $AC$ respectively, with $AX = \frac{1}{3}AB$ and $AY = \frac{1}{3}AC$.

i) Express each of the following in terms of $\mathbf{b}$ and $\mathbf{c}$.

$\vec{BC}$  $\vec{AX}$  $\vec{AY}$  $\vec{XY}$

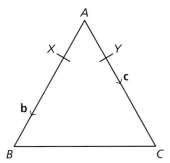

ii) What can you say about $XY$ and $BC$?

# Exercise 12B

1 Which of the following statements refer to vectors?
   i) The speed of light is 186 000 miles per second.
   ii) The spaceship flies to Betelgeuse at $\frac{2}{3}$ the speed of light.

2 A translation has vector $\begin{pmatrix} 5 \\ -8 \end{pmatrix}$. What point is taken to $(3, -2)$ by this translation?

3 Let $\mathbf{u} = \begin{pmatrix} -2 \\ -9 \end{pmatrix}$ and $\mathbf{v} = \begin{pmatrix} -4 \\ 6 \end{pmatrix}$. Find $4\mathbf{u} + 3\mathbf{v}$.

4 With $\mathbf{u}$ and $\mathbf{v}$ as in question 3, illustrate $\mathbf{u}$, $\mathbf{v}$ and $\mathbf{u} - \mathbf{v}$.

5 With $\mathbf{u}$ and $\mathbf{v}$ as in question 3, solve the equation $x\mathbf{u} + y\mathbf{v} = \begin{pmatrix} 2 \\ 33 \end{pmatrix}$.

6 Find the magnitude of the vector $\mathbf{u}$ of question 3.

7 Let $\mathbf{u}$ have magnitude 13 units and direction 220°, and let $\mathbf{v}$ have magnitude 17 units and direction 285°. Find the magnitude and direction of $\mathbf{u} - \mathbf{v}$.

8 Find $m$, given that $(2m, m)$, $(4, 7)$ and $(8, 15)$ are in a straight line.

9 Four points have position vectors $\mathbf{a} + 2\mathbf{b}$, $3\mathbf{a} + 3\mathbf{b}$, $5\mathbf{a} + 6\mathbf{b}$ and $\mathbf{b} - \mathbf{a}$. Which three of the points are in a straight line?

10 $ABCD$ is a quadrilateral. The position vectors of $A$, $B$, $C$ and $D$ are $\mathbf{a}$, $\mathbf{b}$, $\mathbf{c}$ and $\mathbf{d}$ respectively. $X$ and $Y$ are the midpoints of $BD$ and $AC$ respectively.

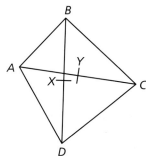

i) Find the position vectors of $X$ and $Y$.
ii) Show that $\vec{AB} + \vec{AD} + \vec{CB} + \vec{CD} = 4\vec{YX}$.

## Exercise 12C

We have described our vectors in terms of coordinate axes which are perpendicular to each other. This isn't essential. The diagram below shows isometric paper, in which the axes are at 60° to each other. Let $\overrightarrow{OA} = \mathbf{a}$ and $\overrightarrow{OB} = \mathbf{b}$.

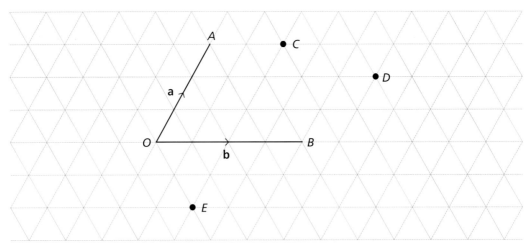

i) Express each of the following in terms of **a** and **b**.
$\overrightarrow{OC}$  $\overrightarrow{OD}$  $\overrightarrow{ED}$

ii) Show that, for any point $X$, $\overrightarrow{OX}$ can be found in terms of **a** and **b**.

## Exercise 12D

So far you have added and subtracted vectors. Multiplication of two vectors may be defined in different ways. One of these ways is called the scalar product or dot product. It is defined by

$$\begin{pmatrix} a \\ b \end{pmatrix} \cdot \begin{pmatrix} c \\ d \end{pmatrix} = ac + bd$$

> Multiply corresponding terms, then add.

**1** Find i) $\begin{pmatrix} 3 \\ 7 \end{pmatrix} \cdot \begin{pmatrix} 8 \\ 2 \end{pmatrix}$   ii) $\begin{pmatrix} 3 \\ -1 \end{pmatrix} \cdot \begin{pmatrix} 2 \\ 5 \end{pmatrix}$

**2** Show that, for any vector **u**, $|\mathbf{u}|^2 = \mathbf{u}.\mathbf{u}$.

**3** Use a result of chapter 7 to show that $\mathbf{u}.\mathbf{v} = 0$ if **u** and **v** are perpendicular.

**4** In fact, $\mathbf{u}.\mathbf{v} = |\mathbf{u}||\mathbf{v}| \cos \theta$, where $\theta$ is the angle between **u** and **v**.
Verify this for $\mathbf{u} = \begin{pmatrix} 3 \\ 2 \end{pmatrix}$ and $\mathbf{v} = \begin{pmatrix} 4 \\ 1 \end{pmatrix}$.

> To find $\theta$, find the angles that the vectors make with the $x$-axis, then subtract.

# Exercise 12E

All the vectors of this chapter have been two dimensional. We can describe translations in three dimensions by three-dimensional vectors, such as $\begin{pmatrix} 2 \\ 3 \\ 6 \end{pmatrix}$.

The solid equivalent of a parallelogram is called a parallelepiped. It has six faces which are parallelograms. Suppose one vertex of a parallelepiped is at $O(0, 0, 0)$, and the three edges leading from $O$ are $\overrightarrow{OA} = \begin{pmatrix} 1 \\ 2 \\ 7 \end{pmatrix}$, $\overrightarrow{OB} = \begin{pmatrix} 2 \\ 8 \\ 1 \end{pmatrix}$ and $\overrightarrow{OD} = \begin{pmatrix} 9 \\ 1 \\ 2 \end{pmatrix}$.

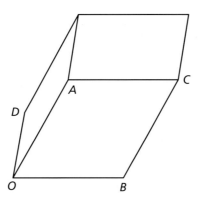

1. Write down the (three-dimensional) coordinates of $A$, $B$ and $C$.
2. Find the coordinates of the other vertices of the parallelepiped.
3. Sketch the parallelepiped.
4. Suppose all the faces of a parallelepiped are rectangles. What is the usual name for the solid?

# 13 Rational and irrational numbers

## Rational numbers

The simplest sort of numbers, names for which exist in every language, are the positive whole numbers or counting numbers. The next simplest are **fractions**, formed by dividing one whole number by another. Another word for a fraction is a **rational** number, because it is the ratio of one whole number to another. These numbers can be positive or negative, the only restriction being that the denominator of the fraction cannot be zero. The following are all rational numbers.

$$\tfrac{3}{4} \qquad 1\tfrac{2}{3} \text{ (i.e. } \tfrac{5}{3}\text{)} \qquad 0.37 \text{ (i.e. } \tfrac{37}{100}\text{)} \qquad -\tfrac{2}{7} \text{ (i.e. } \tfrac{-2}{7}\text{)}$$
$$0.333333\ldots \text{ (this is } \tfrac{1}{3}\text{)} \qquad 0 \text{ (i.e. } \tfrac{0}{1}\text{)}$$

This is the full definition:

- A rational number is any number of the form $\dfrac{a}{b}$, where a and b are whole numbers and $b \neq 0$.

Note that, by this definition, whole numbers are also rational. We can write 17 as $\tfrac{17}{1}$, so 17 is a rational number. This is an example of the **inclusiveness** of mathematical language.

We can do basic arithmetic on rational numbers, and the result is still a rational number.

---

*Example*  Show that the sum of any two rational numbers is also rational.

Let the numbers be $\dfrac{a}{b}$ and $\dfrac{c}{d}$, where $a$, $b$, $c$ and $d$ are whole numbers, and $b$ and $d$ are not zero. Add them

$$\frac{a}{b} + \frac{c}{d} = \frac{ad}{bd} + \frac{bc}{bd} = \frac{ad+bc}{bd}$$

Note that $ad + bc$ and $bd$ are both whole numbers. Also, $bd \neq 0$, as neither $b$ nor $d$ is zero.

So the sum of $\dfrac{a}{b}$ and $\dfrac{c}{d}$ is also a rational number.

---

**Note.** This confirms we were right to allow the word 'rational' to include whole numbers. The sum of $\tfrac{1}{4}$ and $\tfrac{3}{4}$, for example, is a whole number which is not a fraction.

# Exercise 13.1

**1** Show that the following numbers are rational, by writing them in the form $\frac{a}{b}$.

$1\frac{1}{2}$    $0.7$    $1.2$    $3\frac{2}{3}$    $-\frac{3}{5}$
$-1.7$    $3$    $-8$    $-4\frac{1}{5}$    $-3\frac{3}{7}$

**2** Show that the difference of two rational numbers is also rational.
(Hint: as above, let the numbers be $\frac{a}{b}$ and $\frac{c}{d}$.)

**3** Show that the product of two rational numbers is also rational.

**4** Show that when one rational number is divided by another the result is rational, provided that the second number is not zero.

## Terminating and recurring decimals

A decimal which ends is **terminating**. Every terminating decimal is rational. For example:

$$0.5 = \tfrac{1}{2} \quad 0.3 = \tfrac{3}{10} \quad 0.47 = \tfrac{47}{100} \quad 0.28453 = \tfrac{28\,453}{100\,000}$$

A decimal which has a repeated pattern is **recurring**. You are probably familiar with these:

$$\tfrac{1}{3} = 0.333\ldots \quad \tfrac{2}{3} = 0.666\ldots$$

To show that a decimal is recurring, we put a dot or dots over the pattern that recurs. For example

$$\tfrac{1}{3} = 0.333\ldots = 0.\dot{3} \quad\quad 0.23232323\ldots = 0.\dot{2}\dot{3}$$
$$0.544444\ldots = 0.5\dot{4} \quad\quad 0.6032032032\ldots = 0.6\dot{0}3\dot{2}$$

In fact, any recurring decimal is rational. For example

$$0.\dot{5} = 0.555\ldots = \tfrac{5}{9} \quad\quad 0.\dot{4}\dot{5} = 0.454545\ldots = \tfrac{5}{11}$$

To write a recurring decimal as a fraction, we multiply it by the appropriate power of 10.

Take the first example above. Let $x = 0.555\ldots$ Multiply by 10, to shift all the 5s one step along.

$$10x = 5.555\ldots$$

Now we subtract, and all the 5s after the decimal point cancel.

$$10x - x = 5.555\ldots - 0.555\ldots$$
$$9x = 5$$
$$x = \tfrac{5}{9}$$

For $0.\dot{4}\dot{5}$, multiply by 100. Let $x = 0.454545\ldots$

$$100x = 45.454545\ldots$$
$$100x - x = 45.454545\ldots - 0.454545\ldots$$
$$99x = 45$$

Hence $x = \tfrac{45}{99} = \tfrac{5}{11}$.

# RATIONAL AND IRRATIONAL NUMBERS

## Exercise 13.2

1. Write the following recurring decimals using the dot notation.
   **a** 0.777...   **b** 0.12121212...   **c** 0.373737...   **d** 0.128128128...
2. Write out the following recurring decimals to nine decimal places.
   **a** $0.\dot{8}\dot{1}$   **b** $0.7\dot{2}$   **c** $0.58\dot{3}$   **d** $0.\dot{2}0\dot{4}$
3. Enter $\frac{1}{9}$ into your calculator, i.e. enter $1 \div 9$. Write it as a recurring decimal.
4. Write the following as rational numbers:
   **a** 0.444...   **b** 0.777...
5. Enter the decimal 0.52525252 into your calculator. Multiply by 99. You should find that this is very close to an integer. Divide this integer by 99. Do you get 0.52? Hence write 0.52 as a rational number.
6. By the method of question 5, write these as rational numbers.
   **a** $0.\dot{8}\dot{2}$   **b** $0.0\dot{7}$   **c** $0.\dot{3}\dot{6}$   **d** $0.\dot{2}\dot{1}$
7. Consider the recurring decimal $0.\dot{1}2\dot{3}$. Enter 0.123123123 into your calculator, and multiply by 999. Hence write 0.123 as a rational number.
8. Write the following as rational numbers.
   **a** $0.\dot{2}1\dot{8}$   **b** $0.\dot{2}0\dot{7}$   **c** $0.0\dot{5}\dot{4}$   **d** $0.5\dot{4}$   **e** $0.2\dot{7}$   **f** $0.93\dot{5}$
   **g** $0.4\dot{1}\dot{2}$   **h** $0.400\dot{3}$   **i** $0.23\dot{1}\dot{4}$

## Irrational numbers

Not every number is rational. The square root of 2, $\sqrt{2}$, is **irrational**. That is, $\sqrt{2}$ cannot be written as $\frac{a}{b}$, where $a$ and $b$ are whole numbers.

> This was discovered in about 550 BC, by one of the followers of Pythagoras. One story is that the discoverer was drowned for his impiety! Until then, all numbers were believed to be either whole numbers or fractions.

**Proof**
Suppose that we could write $\sqrt{2}$ as $\frac{a}{b}$, where $a$ and $b$ are whole numbers, $b \neq 0$.
If $a$ and $b$ have a common factor, we can cancel it.
   We make the following assumption. We will later show that this assumption is false. We put a * by the side of it as we will refer to it later.

*  $\sqrt{2} = \frac{a}{b}$, where $a$ and $b$ have no common factor

Multiply across by $b$.
$$\sqrt{2}b = a$$
$$2b^2 = a^2 \qquad [1]$$

Hence $a^2$ is even.
   Now, when an odd number is multiplied by itself, the result is also odd. In particular, the square of an odd number is odd. So, as $a^2$ is even, $a$ itself cannot be odd.
   Hence $a$ is even. Write $a$ as $2c$, and substitute into the equation [1] above
$$2b^2 = (2c)^2 = 4c^2$$
$$b^2 = 2c^2$$

Hence $b^2$ is even. Hence, as above, $b$ is even.
   Therefore $a$ and $b$ are both even. So they have the common factor 2. This is a contradiction, as the assumption * states that $a$ and $b$ have no common factor.

# Irrational numbers

The original assumption *, that we can write $\sqrt{2}$ as $\frac{a}{b}$, where $a$ and $b$ are whole numbers, must be false. $\sqrt{2}$ is not a rational number.

Therefore $\sqrt{2}$ is an irrational number.

**Note.** The method of proof here is called **proof by contradiction**. We assumed that something was true (that $\sqrt{2}$ is $\frac{a}{b}$, where $a$ and $b$ are whole numbers) and obtained a contradiction. Hence the original assumption is false.

Many other numbers are irrational. For example, $\sqrt{3}$, $\sqrt{5}$, $\sqrt{6}$ and so on are irrational. The square root of any whole number which is not a perfect square is irrational. The number $\pi$ is irrational, though it is very hard to prove this. The approximations of $\frac{22}{7}$ or 3.142 for $\pi$ are only approximations – we cannot write down the exact value of $\pi$ as a fraction or a decimal.

When you combine irrational numbers, the result may be irrational.

*Example*  Show that $3 + \sqrt{2}$ is irrational.

Suppose that $3 + \sqrt{2}$ was rational. Then by the results above, $3 + \sqrt{2} - 3$ is also rational.
But $3 + \sqrt{2} - 3 = \sqrt{2}$, which we known is irrational. Hence $3 + \sqrt{2}$ is irrational.

## Exercise 13.3

1. Show that $2 + \sqrt{2}$ is irrational.
2. Show that $5 - \sqrt{2}$ is irrational.
3. Show that $5\sqrt{2}$ is irrational.
4. Show that $7\sqrt{2}$ is irrational.
5. Show that $\frac{1}{6}\sqrt{2}$ is irrational.
6. Show that $\frac{2}{7}\sqrt{2}$ is irrational.

The sum of two irrationals *may* be irrational. For example, $\sqrt{2} + \sqrt{3}$ is irrational. But sometimes the sum of two irrationals is rational.

*Examples*  Find two different irrational numbers whose sum is rational.

We know that $\sqrt{2}$ is irrational. Subtract $\sqrt{2}$ from 2, and the result is still irrational. But the sum of these is 2, which is rational.

$\sqrt{2}$ and $2 - \sqrt{2}$ are irrational. But $\sqrt{2} + (2 - \sqrt{2}) = 2$, which is rational.

Find two different irrational numbers whose product is rational.

We know that $\sqrt{2}$ is irrational. So also is $3\sqrt{2}$. But their product is

$$\sqrt{2} \times 3\sqrt{2} = 3 \times (\sqrt{2})^2 = 3 \times 2 = 6$$

$\sqrt{2}$ and $3\sqrt{2}$ are irrational, but $\sqrt{2} \times 3\sqrt{2}$ is rational.

# 212 RATIONAL AND IRRATIONAL NUMBERS

## Exercise 13.4

1. Find another pair of irrational numbers whose sum is rational.
2. Find another pair of different irrational numbers whose product is rational.
3. Find two different irrational numbers whose difference is rational.
4. Find two different irrational numbers whose ratio is rational.
5. From the numbers below, find
   a. a pair whose sum is rational
   b. a pair whose sum is irrational
   c. a pair whose product is rational
   d. a pair whose ratio is irrational.

   $2 + \sqrt{3} \quad 6 + \sqrt{3} \quad 5\sqrt{3} \quad 4 + 2\sqrt{3} \quad 7 - \sqrt{3} \quad \sqrt{27}$

6. Find an irrational number between 10 and 12.
7. Find a rational number between $\sqrt{2}$ and $\sqrt{3}$.
8. Find an irrational number between $\frac{1}{2}$ and $\frac{1}{3}$.
9. Find a rational number between $\sqrt{0.2}$ and $\sqrt{0.3}$.

## Arithmetic of surds

A **surd** is any expression involving roots of rational numbers. Examples of surds are

$$2 + \sqrt{3} \quad 7 - \sqrt{2} \quad 2\sqrt{5} + 4\sqrt{7} \quad \tfrac{1}{2}\sqrt{\tfrac{3}{4}}$$

When surds are combined by arithmetical operations, the results can sometimes be simplified. Note that for example:

$$\sqrt{2} \times \sqrt{2} = 2 \quad \sqrt{2} \times \sqrt{3} = \sqrt{6} \quad 3\sqrt{2} + 5\sqrt{2} = 8\sqrt{2}$$

**Note.** When we multiply the square roots of two numbers, the result is the square root of the product.

$$\sqrt{a} \times \sqrt{b} = \sqrt{ab}$$

There is no way to simplify the *sum* of two square roots.

$$\sqrt{a} + \sqrt{b} \neq \sqrt{a+b}$$

Just try any pair of non-zero numbers. For example $\sqrt{2} + \sqrt{3} \neq \sqrt{5}$.

*Examples* Simplify $\sqrt{3} + \sqrt{12}$.

Note that
$$\sqrt{12} = \sqrt{(4 \times 3)}$$
$$= \sqrt{4} \times \sqrt{3}$$
$$= 2 \times \sqrt{3}$$

Hence $\sqrt{3} + \sqrt{12} = \sqrt{3} + 2\sqrt{3}$.
Therefore $\sqrt{3} + \sqrt{12} = 3\sqrt{3}$.

# Arithmetic of surds

Expand and simplify $(2+\sqrt{3})(3-\sqrt{3})$.

Multiply both terms in the first bracket by both terms in the second bracket.
$$(2+\sqrt{3})(3-\sqrt{3}) = 2\times 3 - 2\times\sqrt{3} + 3\times\sqrt{3} - \sqrt{3}\times\sqrt{3}$$
Note that $-2\times\sqrt{3} + 3\times\sqrt{3} = 1\times\sqrt{3}$ and that $\sqrt{3}\times\sqrt{3} = 3$.
$$(2+\sqrt{3})(3-\sqrt{3}) = 6 + 1\times\sqrt{3} - 3$$
Hence $(2+\sqrt{3})(3-\sqrt{3}) = 3 + \sqrt{3}$.

## Exercise 13.5

**1** Simplify these expressions.
  **a** $\sqrt{3}\times\sqrt{3}$
  **b** $\sqrt{5}\times\sqrt{5}$
  **c** $\sqrt{3}+\sqrt{3}$
  **d** $\sqrt{2}+\sqrt{2}$
  **e** $3\sqrt{5}-\sqrt{5}$
  **f** $4\sqrt{7}+3\sqrt{7}$

**2** Simplify these expressions.
  **a** $\sqrt{2}+\sqrt{8}$
  **b** $\sqrt{7}+\sqrt{63}$
  **c** $\sqrt{20}+\sqrt{45}$
  **d** $3\sqrt{2}-4\sqrt{8}$
  **e** $5\sqrt{10}-\sqrt{40}$
  **f** $\sqrt{28}-\sqrt{7}$

**3** Expand these expressions and simplify as far as possible.
  **a** $(3+\sqrt{2})(1+\sqrt{2})$
  **b** $(2-\sqrt{2})(3+\sqrt{2})$
  **c** $(5+\sqrt{5})(3-\sqrt{5})$
  **d** $(1+2\sqrt{3})(4-3\sqrt{3})$
  **e** $(3+3\sqrt{2})(5-2\sqrt{2})$
  **f** $(3-2\sqrt{5})(4-3\sqrt{5})$

Suppose the denominator of a fraction is a square root. Then the fraction is sometimes simplified if we multiply numerator and denominator by the same square root. For example, $\dfrac{4}{\sqrt{2}}$ is such a fraction. Multiply numerator and denominator by $\sqrt{2}$.

$$\frac{4}{\sqrt{2}} = \frac{4}{\sqrt{2}} \times \frac{\sqrt{2}}{\sqrt{2}} = \frac{4\sqrt{2}}{\sqrt{2}\times\sqrt{2}} = \frac{4\sqrt{2}}{2} = 2\sqrt{2}.$$

*Example* Simplify $\sqrt{3} + \dfrac{2}{\sqrt{3}}$.

Multiply numerator and denominator of $\dfrac{2}{\sqrt{3}}$ by $\sqrt{3}$.

$$\frac{2}{\sqrt{3}} \times \frac{\sqrt{3}}{\sqrt{3}} = \frac{2\sqrt{3}}{\sqrt{3}\times\sqrt{3}} = \frac{2\sqrt{3}}{3} = \tfrac{2}{3}\sqrt{3}$$

Hence $\sqrt{3} + \dfrac{2}{\sqrt{3}} = \sqrt{3} + \tfrac{2}{3}\sqrt{3}$.

$$\sqrt{3} + \frac{2}{\sqrt{3}} = 1\tfrac{2}{3}\sqrt{3}.$$

# 214 RATIONAL AND IRRATIONAL NUMBERS

## Exercise 13.6

**1** Simplify the following expressions.

a $\dfrac{6}{\sqrt{2}}$  

b $\dfrac{6}{\sqrt{3}}$  

c $\dfrac{15}{\sqrt{5}}$  

d $3\sqrt{2} - \dfrac{4}{\sqrt{2}}$  

e $\dfrac{6}{\sqrt{3}} - \sqrt{3}$  

f $\dfrac{10}{\sqrt{5}} + 3\sqrt{5}$  

g $\dfrac{5}{\sqrt{2}} + \dfrac{7}{\sqrt{2}}$  

h $\dfrac{2}{\sqrt{7}} + \dfrac{5}{\sqrt{7}}$  

i $3\sqrt{10} - \dfrac{\sqrt{2}}{\sqrt{5}}$  

j $\dfrac{\sqrt{3}}{\sqrt{2}} + \sqrt{6}$  

k $\sqrt{21} + \dfrac{4\sqrt{7}}{\sqrt{3}}$  

l $\dfrac{\sqrt{3}}{\sqrt{2}} + \dfrac{\sqrt{2}}{\sqrt{3}}$

**2 a** Show that $(\sqrt{2}-1)(\sqrt{2}+1)$ is rational.

**b** Multiply the numerator and denominator of $\dfrac{2}{\sqrt{2}-1}$ by $(\sqrt{2}+1)$. Hence simplify the expression.

**3 a** Show that $(3-\sqrt{5})(3+\sqrt{5})$ is rational.

**b** Multiply the numerator and denominator of $\dfrac{8}{3+\sqrt{5}}$ by $(3-\sqrt{5})$. Hence simplify the expression.

**4** By the method of questions 2 and 3, simplify these expressions.

a $\dfrac{2}{\sqrt{3}-1}$  

b $\dfrac{1}{2-\sqrt{3}}$  

c $\dfrac{1}{\sqrt{3}-\sqrt{2}}$

## Use of surds

Surds arise in many mathematical contexts, in particular when using Pythagoras' theorem or in trigonometry. It is often more accurate to leave a result as a surd rather than to evaluate it to a certain number of significant figures (unless of course you are asked to give the answer to a certain number of significant figures).

*Examples* Find the unknown length $x$ in the diagram shown.

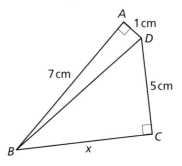

First find the hypotenuse $BD$ of the top triangle. By Pythagoras' theorem, it is equal to

$$\sqrt{7^2 + 1^2} = \sqrt{50}$$

Leave this as $\sqrt{50}$. Apply Pythagoras' theorem in the right-hand triangle.

$$x = \sqrt{(\sqrt{50})^2 - 5^2} = \sqrt{50 - 25} = \sqrt{25} = 5$$

The unknown side has length exactly 5 cm.

**Note.** If we had evaluated $\sqrt{50}$ as 7.07, and then squared this, we would have obtained

$$\sqrt{7.07^2 - 5^2} = \sqrt{49.98 - 25} = \sqrt{24.98} = 4.998$$

Though this is very close to 5, it is not exactly so. The final answer should be exactly 5 cm.

The diagram shows an isosceles right-angled triangle, with the shorter sides equal to 1 unit. Use the triangle to find the exact value of $\sin 45°$.

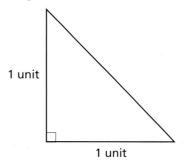

The OPP and ADJ of the triangle are both equal to 1. The HYP side is found by Pythagoras' theorem

$$\text{HYP} = \sqrt{1^2 + 1^2} = \sqrt{2}$$

The smaller angles of the triangle are each 45°.

Hence the exact value of $\sin 45°$ is $\dfrac{1}{\sqrt{2}}$.

## Exercise 13.7

**1** Find the unknown lengths in these diagrams. Where appropriate, leave your answer as a surd.

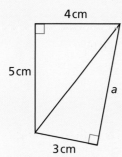
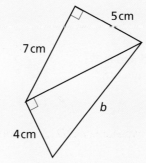
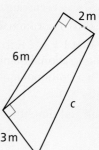

**2** Use the isosceles right-angled triangle with shorter sides 1 unit, in the second example above, to find the exact value of $\cos 45°$.

# 216 RATIONAL AND IRRATIONAL NUMBERS

**3** The diagram shows an equilateral triangle $ABC$ of side 2 units. The midpoint of $BC$ is $D$.

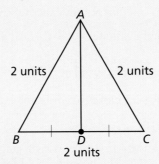

**a** Write down the angles of the triangle.
**b** Find the length of $AD$, leaving your answer as a surd.
**c** Find the exact values of the following, leaving your answers as surds where relevant.
   i) $\sin 60°$   ii) $\cos 60°$   iii) $\tan 30°$   iv) $\tan 60°$

## SUMMARY

- A number that can be written as $\dfrac{a}{b}$, where $a$ and $b$ are integers and $b \neq 0$, is a **rational** number.
- The sum, difference, product and ratio of rational numbers are also rational (provided that we are not dividing by 0).
- Every **terminating** decimal and every **recurring** decimal is rational.
- A number that is not rational is **irrational**. Many numbers such as $\sqrt{2}$, $\sqrt{3}$ and $\pi$ are irrational.
- An expression involving roots is a **surd**. A surd expression can often be simplified.

## Exercise 13A

1 Show that $5\frac{3}{7}$ is rational by writing it as $\dfrac{a}{b}$.
2 Write $\frac{2}{11}$ as a recurring decimal.
3 Write $0.1\dot{7}$ as a rational.
4 Show that $1 + \sqrt{2}$ is irrational.
5 Find an irrational number between 5 and 6.
6 Find two irrational numbers whose sum is 8.
7 Simplify $\sqrt{5} \times \sqrt{125}$.
8 Expand and simplify $(3 - 2\sqrt{2})(5 - 3\sqrt{2})$.

9 Simplify $\sqrt{3} + \dfrac{7}{\sqrt{3}}$.

10 Find the length $x$ in the diagram shown. Leave your answer as a surd.

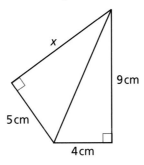

## Exercise 13B

1 Suppose $x$ is a rational number $\left(\text{so } x = \dfrac{a}{b} \text{ for some integers } a \text{ and } b\right)$. Show that $1 + x$ is a rational number.
2 Write $\tfrac{7}{9}$ as a recurring decimal.
3 Write $0.2\dot{3}$ as a rational.
4 Show that $5\sqrt{3}$ is irrational.
5 Find a rational number between $\sqrt{7}$ and $\sqrt{5}$.
6 Find two irrational numbers whose product is 9.
7 Simplify $3\sqrt{7} - \sqrt{343}$.
8 Expand and simplify $(2 - 5\sqrt{2})(2 - \sqrt{2})$.
9 Simplify $\dfrac{\sqrt{2}}{\sqrt{5}} + \dfrac{\sqrt{5}}{\sqrt{2}}$.
10 Find the length $x$ in the diagram shown. Leave your answer as a surd.

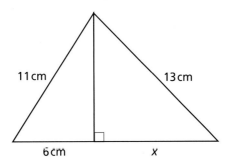

## Exercise 13C

In this chapter we proved that $\sqrt{2}$ is irrational. Amend the proof to show that $\sqrt{3}$ is irrational. Why doesn't the proof work for $\sqrt{4}$ (which is 2)?

## Exercise 13D

Above we showed that the exact value of $\sin 45°$ is $\dfrac{1}{\sqrt{2}}$. In the exercise that followed you were asked to find the exact values of $\sin 60°$ etc. Many other angles have sines, cosines and tangents whose exact values are surds. For example

$$\tan 15° = 2 - \sqrt{3}$$

$$\sin 15° = \sqrt{\dfrac{1}{2} - \dfrac{\sqrt{3}}{4}}$$

$$\cos 36° = \tfrac{1}{4}(\sqrt{5} + 1)$$

**1** Use your calculator to check the results above.

**2** Find

  **a** $\tan^{-1}(2+\sqrt{3})$     **b** $\sin^{-1}\sqrt{\dfrac{1}{2} + \dfrac{\sqrt{3}}{4}}$     **c** $\cos^{-1}(\tfrac{1}{4}(\sqrt{5}-1))$

**3** It can be shown that, for all angles $x$:

$$\cos 2x = 2\cos^2 x - 1 \qquad \cos 2x = 1 - 2\sin^2 x \qquad \tan 2x = \dfrac{2\tan x}{1 - \tan^2 x}$$

> Here $\cos^2 x$ means $(\cos x)^2$

Use these results to find the following in surd form.
  **a** $\sin 22.5°$    **b** $\cos 18°$    **c** $\tan 22.5°$

**4** In fact, the sine, cosine and tangent of an angle of a whole number of degrees can be expressed in terms of square roots, provided that the number is divisible by 3. In particular

$$\sin 3° = \sqrt{\dfrac{1}{2} - \dfrac{\sqrt{3}}{16} - \dfrac{\sqrt{15}}{16} - \sqrt{\dfrac{5}{128} - \dfrac{\sqrt{5}}{128}}}$$

Check it!

## Exercise 13E

The following is adapted from *De l'esprit Géométrique* by Blaise Pascal (1623–1662). In it he argues that space is infinitely divisible.

> *If it were true that space is composed of a certain finite number of indivisible parts, then it would follow that for two shapes, both being square, one being double the area of the other, one would contain double the number of indivisible parts of the other. If anyone can arrange points in squares, so that one has double the number of points of the other, then I will put him ahead of all the geometers of the world.*

Suppose the smaller square contains $n$ points. How many points are there in the larger square? By considering the sides of the squares, use a result of this chapter to show that the larger square cannot contain a whole number of points.

# Making and spending money

Everyone uses mathematics in the context of money. Working out how much we should be paid, planning the purchase of an expensive item, calculating how much we have to pay in tax; all these involve mathematics.

## Making money

There are several ways you can make money. You could receive a **wage**, which is usually paid weekly. Your income could be given as an annual **salary**, which is usually paid monthly. If you are self-employed, you might be paid for each bit of work that you do. If you are an author or a musician, you might be paid royalties for each book or each recording of yours that is sold.

### Overtime

In most places of work, there is a fixed number of hours per day. When you are asked to work longer hours, you are paid **overtime**. This might be twice the hourly rate. It might be 'time and a half', which is $1\frac{1}{2}$ times the standard hourly rate.

*Example*  Nick normally works a 38-hour week for £7.20 per hour. One week he works an extra six hours at time and a half. How much does he earn this week?

Nick's ordinary pay is $38 \times £7.20$, which is £273.60.
The overtime hourly rate is $1\frac{1}{2} \times £7.20$, which is £10.80. Multiply this by 6. The overtime pay is $6 \times £10.80$, which is £64.80.

His total earnings are £338.40.

## Exercise M1

1. Lauren is paid at £8 per hour for a 36-hour week. How much does she get for a week in which she works 40 hours, including 2 hours at double time?
2. Jane's job pays £6.80 per hour, and overtime is at double time. What does she earn for a week of 40 hours standard rate and 4 hours overtime?
3. Karim is paid a standard rate of £9.60 per hour, for a week of 40 hours. How much does he receive for a week in which he worked a total of 44 hours, including 4 hours at time and a half?
4. Michael earns £7.50 per hour for a week of 38 hours. Overtime is at time and a half. In a certain week he earned £375. How many hours of overtime did he work?
5. Su Ying works a 36-hour week, and overtime is at time and a half. She was paid £332.10 for a week of 39 hours. What is her basic hourly rate?
6. Self-employed people often get their income from many different sources. An author, for example, doesn't just get money from the sale of books.
   a. Royalties are paid at a rate of 10% on the price of the books sold. How much does the author get if 15 600 books are sold at £10.80 each?
   b. A fee is payable to authors when their books are photocopied. If 62 000 pages are photocopied at 5.25p each, how much does the author get?
   c. An author received £1491 in photocopying fees. Use the rate in part **b** to find how many pages were copied.
   d. **Public lending right** pays authors 2.49p for each time one of their books is borrowed from a public library. How much does an author get if his books are borrowed 18 000 times?
   e. An author received £557.76 from public lending right. How many times were her books borrowed?

# MAKING AND SPENDING MONEY

## Profit and loss

Frequently we buy something and then sell it later. If we gain money it is a profit; if we lose money it is a loss. Often the profit or loss is expressed as a percentage of the buying price.

*Example*  Mr Simons bought a car from a garage for £4000.
**a** He later sold the car for £3600. What was his percentage loss?
**b** The garage made a profit of 25% when it sold the car to Mr Simons. What price had the garage bought the car for?

**a** The loss is £400. Express this as a percentage of the buying price.
$$\frac{400}{4000} \times 100 = 10$$
He made a loss of 10%.

**b** When the garage made a profit of 25%, it increased the original buying price by a factor of 1.25. So to go from the selling price to the buying price, *divide* by 1.25.
$$4000 \div 1.25 = 3200$$
The garage had bought the car for £3200.

**Note.** Throughout this chapter we shall have many problems about increasing or decreasing by a certain percentage. It is always best to solve these problems by a multiplying factor, as in part **b** above. For example, to increase by 15%, multiply by 1.15. To decrease by 15%, multiply by 0.85.

## Exercise M2

1. A chair is bought for £240 and sold for £300. What is the percentage profit?
2. A picture is bought for £12 500 and sold for £15 000. What is the percentage profit?
3. A bicycle is bought for £120 and sold for £100. What is the percentage loss?
4. Jason bought a car for £3400 and sold it at a profit of 15%. What price did he sell it for?
5. Mary bought a computer for £1200 and sold it a loss of 28%. What price did she sell it for?
6. A house is sold for £88 200, making a profit of 5%. What price was it bought for?
7. A market trader sold a batch of crockery for £759, making a profit of 38%. What price did he pay for the crockery?
8. A computer is sold for £935, making a loss of 15%. What price was it bought for?
9. A car is sold for £4680, making a loss of 35%. What price was it bought for?

## Interest

When you invest money in a bank or building society, it earns **interest**. If you take the interest out, so that the original sum is unchanged, this is **simple interest**. If you leave the money to accumulate, then this is **compound interest**.

*Example*  £2000 is invested at 8% compound interest. How much is there after four years?

Every year the money increases by 8%, so it is multiplied by 1.08. After four years the money has been multiplied by 1.08 four times, i.e. by $1.08^4$.
$$2000 \times 1.08^4 = 2721$$
After four years the amount is £2721.

## Exercise M3

1. £3000 is invested at 6% compound interest. How much is there after three years?
2. £4500 is invested at 7% compound interest. How much is there after five years?
3. £8000 was invested at compound interest, and after two years it had increased to £9504.80. What was the rate of interest?
4. £1600 was invested at compound interest, and after three years it had increased to £2129.60. What was the rate of interest?
5. After two years at 6% compound interest, a sum of money had increased to £5618. What was the original sum?
6. The value of a car decreases by 12% each year. If it is bought for £11 500, how much is it worth after three years?
7. The value of a computer decreases by 28% each year. If it is bought for £980, how much is it worth after three years?

## Spending money

The whole point of making money is to spend it. Sometimes you pay for what you want immediately. For example, when you buy a can of drink, you have to pay for it at the time. But sometimes for more expensive items you spread the payment over a longer time. This is known as **hire purchase** (HP) or the 'Never-never'.

Often there is an initial sum (or down payment), then several monthly or weekly payments.

*Because it seems that you will never stop paying for it!*

*Example*    A computer is advertised at £100, plus £25 per month for 24 months. What is the total cost of buying it?

Multiply £25 by 24. The result is £600. Add this on to £100.

The total cost is £700.

## Exercise M4

In questions 1–3, find the total cost of buying the item.

1. A motorcycle, costing £150 down payment and £50 per month for 12 months
2. A sofa, costing £200 down payment and £20 per month for 12 months
3. A car, costing £2000 down payment and £300 per month for 24 months
4. Suppose the motorcycle of question 1 can be bought outright for £650. What is the extra percentage cost in buying it by HP?
5. Suppose the car in question 3 can be bought outright for £7500. What is the extra percentage cost in buying it by HP?

### Foreign currency

When you travel abroad, you usually have to convert your money into the local money. The conversion rate between the two currencies is the **rate of exchange**.

*Example* At a certain time the rate of exchange between the £ and the $ is $1.46 per £.
a Convert £830 to dollars.
b Convert $1934 to pounds.

a Multiply 830 by 1.46.
 £830 is worth $1211.80.
b Divide $1934 by 1.46.
 $1934 is worth £1324.66  (to the nearest 1p).

## Exercise M5

1. Use the £/$ rate above to convert
   a £2880 to $
   b $2377 to £
2. The exchange rate of the euro to the pound is €1 per £0.58.
   a Convert £155 to €.
   b Convert €744 to £.
3. At a certain time 1200 South African rand are worth £263.
   a What is the exchange rate, in rand per £?
   b At this rate, how many rand is £400 worth?
4. The rates at which a bank buys and sells foreign currency differ. Suppose the buying rate for $ to £ is $1.483 per £ (i.e. the bank will take $1.483 and give you £1) and the selling rate is $1.475 per £ (i.e. the bank will take £1 and give you $1.475).
   Mrs Snyder changes $1840 to £, but then has to change it back to $. How much has she lost?
5. Suppose the buying rate for the Japanese yen is ¥174 per £, and the selling rate is ¥168 per £. Mr Okada comes to the UK with ¥200 000 and changes it to £. He spends £643, then changes the rest back to ¥. How much does he get?

## Spending money

### Borrowing

You can borrow money to buy something which you cannot afford at the moment and then pay it back later, with interest. In particular, most people buy their homes by borrowing money through a **mortgage**.

If you don't pay a credit card bill, then you are charged interest on the amount outstanding (left unpaid). This is charged monthly, and there is an equivalent annual rate, called an APR (annual percentage rate).

*Examples* A mortgage of £30 000 is taken out. At the beginning of each year, interest at 8% is charged on the amount outstanding, and at the end of the year a repayment of £3200 is made. Find out how much is outstanding at the end of the first three years.

First increase £30 000 by 8%, then subtract £3200.

$$30\,000 \times 1.08 - 3200 = 29\,200$$

After one year, £29 200 is outstanding.

Now increase this by 8%, then subtract £3200. After two years, £28 336 is outstanding. Repeat for the third year.

After three years, £27 402.88 is outstanding.

Find the APR for a monthly interest rate of 2%.

Every month the amount outstanding is multiplied by 1.02. After 12 months it is multiplied by $1.02^{12}$. This is equal to 1.268.

The APR is 26.8%.

**Note.** This is a very high rate of interest. It is unwise to borrow money on a credit card. Also note that the APR is greater than $12 \times 2\%$, i.e. 24%.

## Exercise M6

1 A mortgage of £40 000 at 7% interest is taken out. The annual repayments are £3500. Find how much is still owing at the end of each of the first three years.
2 A mortgage of £35 000 at 8% interest is taken out. The annual repayments are £3200. Find how much is outstanding at the end of each of the first three years.
3 A mortgage of £40 000 is taken out at 7% interest. After the first annual repayment the amount outstanding is £39 500. What was the repayment?
4 A mortgage of £45 000 was taken out, and after the first annual repayment of £4000 the amount outstanding was £44 825. What was the rate of interest?
5 A mortgage of £38 000 was taken out, and after the first annual repayment of £3100 the amount outstanding was £37 750. What was the rate of interest?
6 A mortgage of £100 000 was taken out with the rate of interest at 5% for each of the first two years. After the second annual repayment the amount outstanding was £89 750. What was the annual repayment?
7 A mortgage of £100 000 was taken out. After two annual repayments of £10 000 each the amount outstanding was £91 760. What was the annual rate of interest?
8 Find the APR for a monthly interest rate of 1.5%.
9 Find the monthly interest rate which gives an APR of 34.5%.

## Taxes

People earn money, and then they spend it. But the amounts aren't equal. People don't get to keep all the money they earn – the government takes a large proportion of earnings before they can be spent. These are **taxes**.

One tax you have to pay when buying something is **Value Added Tax** (VAT). This is a fixed percentage of the original cost. At present the rate of VAT is $17\frac{1}{2}\%$.

To find $17\frac{1}{2}\%$ of a number, multiply by $17\frac{1}{2}/100$. You might find it easier to use decimals, and multiply by 0.175. To find the cost of an item including VAT, multiply the original cost by 1.175. If you are given the cost including VAT, then to find the original cost *divide* by 1.175.

*Examples*  What is the VAT on a CD costing £8?

Multiply 8 by 0.175: $8 \times 0.175 = 1.4$.

The VAT is £1.40.

The price of a printer including VAT is £305.50. What was the price before VAT?

Divide 305.50 by 1.175.
The price before VAT was £260.

## Exercise M7

In questions 1–3 find the VAT, at a rate of $17\frac{1}{2}\%$.

1  A computer costing £800
2  Building work costing £700
3  A lawyer's fee of £600
4  Suppose the rate of VAT falls to 15%. What is the VAT on a £60 dress?
5  The VAT on electricity bills is at a rate of 5%. An electricity bill is £120 before VAT. How much is the VAT?
6  Suppose the VAT rises to 20%. What is the VAT on a meal costing £30?
7  The VAT rate on electricity bills used to be 8%. What was the VAT on a bill of £150?
8  The VAT rate increases from $17\frac{1}{2}\%$ to 20%. What is the percentage increase in the VAT received by the government? Give your answer correct to the nearest whole number.
9  The following prices include VAT at $17\frac{1}{2}\%$. In each case find the price before VAT.
   a  A computer, for £1151.50
   b  Decoration work, for £470
   c  Do It Yourself supplies, for £211.50
10 The government increases the rate of VAT, and as a result the price of an item increases by 2%. What is the new rate of VAT?

## Income tax

The main source of funds for the government is **income tax**. This is tax levied on the money earned by people. Income tax is meant to be a *progressive* tax. This means that rich people pay a higher proportion of their income than poor people do.

Everyone has a **tax-free allowance**, i.e. an amount on which no tax is paid. So people with low incomes do not pay much tax, as the allowance is a large part of their income. Income above the allowance is taxed at a certain percentage. This percentage increases at higher income levels. This also makes the rich pay a higher proportion of their income in tax.

To work out the tax payable, subtract the tax-free allowance. Income over the allowance is divided into different bands, which are taxed at different percentage rates. The details for a single person (for one year) might be:

| | |
|---|---|
| tax-free allowance | £4535 |
| next £1880 | taxed at 10% |
| next £27 520 | taxed at 22% |
| rest of income | taxed at 40% |

*Examples*  Use the scheme above to find the tax payable on an annual income of £36 095.

First find how much income lies in the different bands. Subtract the allowance of £4535 from the income of £36 095.

$$36\,095 - 4535 = 31\,560$$

The first £1880 is taxed at 10%. Subtract this.

$$31\,560 - 1880 = 29\,680$$

The next £27 520 is taxed at 22%. Subtract this.

$$29\,680 - 27\,520 = 2160$$

So there is £4535 tax free, £1880 at 10%, £27 520 at 22% and the remaining £2160 at 40%. Find all the amounts of tax.

| | |
|---|---|
| £1880 is taxed at 10% | £188 |
| £27 520 is taxed at 22% | £6054.40 |
| £2160 is taxed at 40% | £864 |

Add these amounts.
The tax due is £7106.40.

Ms Corrigan is a single person who pays £4808 in tax annually. What is her income?

From the example above, note that the tax paid at the lowest band is £188. So of this tax, £188 is at the lower rate of 10%.

$$4808 - 188 = 4620$$

From the result above, 22% of £27 520 is £6054.40. This is greater than £4620. Hence she does not pay any tax in the highest band. To find her income in the middle band, divide by 0.22.

$$4620 \div 0.22 = £21\,000$$

So she has £21 000 in the middle band, £1880 in the lowest band and an allowance of £4535. Add these together.
Her income is £27 415.

## 226 MAKING AND SPENDING MONEY

### Exercise M8

1. Find the tax paid by the following people, assuming that they all fit into the tax scheme described on page 225.
   a. Ms Patel, with an income of £20 000
   b. Mr Barden, with an income of £30 000
   c. Mr McNeil, with an income of £18 000
   d. Miss Murdoch, with an income of £46 500
2. Find the incomes of these people, assuming that they all fit into the tax scheme on page 225.
   a. Mr Jones, who pays £1301.20 in tax annually
   b. Ms Prentice, who pays £5380 in tax annually
   c. Miss Deeney, who pays £8850.40 in tax annually
3. The tax rate in the middle band decreases to 20%, and the rate in the upper band increases to 50%. Find the percentage increase or decrease in the tax paid by people with the following annual incomes.
   a. £19 430
   b. £41 800
   c. £34 400
4. The tax changes described in question 3 do not alter the amount of tax that Mr Mahon is paying. What is his income?

### SUMMARY

- Income is paid as **wages** (usually every week) or as **salary** (usually every month).
- Extra hours of work are paid at a higher rate. This **overtime** is sometimes double time, and sometimes time and a half.
- Money that is invested in a bank or building society earns **simple interest** or **compound interest**.
- Goods can be paid for in full, or they can be paid for by an initial sum and regular payments (**hire purchase**).
- The conversion rate between two currencies is the **rate of exchange**.
- Money can be borrowed and paid back later, with interest, as with a **mortgage**.
- **Value Added Tax** (VAT) is a tax on the price of goods. At present it is $17\frac{1}{2}$%.
- **Income tax** is levied on incomes in different bands, as percentages of annual income after a **tax-free allowance** has been subtracted.

### Exercise MA

1. Brian is paid a basic rate of £8.60 per hour for a 38-hour week and overtime is at time and a half. In a certain week he earned £404.20. How much overtime did he do?
2. Dario buys a motorbike for £840 and sells it for £882. What is the percentage profit?
3. £9000 is invested at 8% compound interest. How much is there after three years?
4. A pair of shoes costs £47 including VAT at $17\frac{1}{2}$%. What was the cost before VAT?
5. The buying and selling rates for the euro are €1.54 and €1.48 respectively. Mr Bourrely changes €750 to £, spends £300 and then changes the rest back to euros. How much does he get?
6. Ms Wender changed euros to £ at €1.48 per £ but then had to change the money back to euros. If she made a percentage loss of 1.5%, what was the rate at which she changed back from £ to €?
7. After two years at compound interest, £4000 had increased to £4665.60. What was the rate of interest?
8. Find the APR for a monthly interest rate of 2.3%.
9. Use the scheme on page 225 to find the tax payable on an annual income of £7100.
10. Mrs Flanelly pays £9642.40 annually in income tax. Use the scheme on page 225 to find her income.

## Exercise MB

1. Hugh works a 38-hour week, with a basic pay of £8.60 per hour. In a week in which he worked 42 hours he received £387. What is the overtime rate, as a multiple of the standard rate?
2. A car is bought for £8400 and sold for £7000. What is the percentage loss?
3. After three years of compound interest at 8%, a sum of money has increased to £30 233. What was the original sum?
4. A scanner costs £141, including VAT at $17\frac{1}{2}$%. What is the cost before VAT?
5. A car can be bought outright for £11 500, or for a down payment of £3000 and 36 monthly payments of £300. What is the extra percentage cost in buying by the second method?
6. The buying and selling rate for the Japanese yen are ¥172.3 and ¥168.3 respectively. Izumi changes ¥300 000 to £ for her holiday. At the end of her holiday she changes the remaining £ back to ¥, obtaining ¥91 000. How much did she spend?
7. The value of a computer fell from £1250 to £820 over two years. Find the annual percentage decrease in its value, assuming that this is constant.
8. A sum of money is borrowed at a monthly rate which has an APR of 36%. What is the monthly rate of interest?
9. Two simple income tax schemes are proposed:

    Scheme A: no tax free allowance, first £5000 of income taxed at 10%, income over £5000 taxed at 20%.
    Scheme B: tax free allowance of £10 000, income over £10 000 taxed at 25%.

    Find the amounts paid on an annual income of £28 000 under each of these schemes.
10. Refer to the schemes of question 9. Miss Gillespie finds that she would pay twice the amount of tax under scheme A than under scheme B. What is her annual income?

## Exercise MC

No one likes paying tax, but government services have to be paid for. Various ways of taxing have been discussed in this chapter:

- a tax on spending, i.e. VAT
- a tax on income, i.e. income tax.

1. Other methods of taxing are in use. For example, we have to pay 'duty' on goods like tobacco and petrol. Find some other methods of taxing.
2. Write down what you think is good and what is bad about each method of taxing.

## Exercise MD

At present the rate of VAT is $17\frac{1}{2}$%. What an awkward number!
 Suppose a computer game costs £40 before VAT. How much is the VAT? (Use a calculator.)
 There is a way to find the VAT without a calculator.

| Take a tenth of £40 | £4 |
| Take half of this amount | £2 |
| Take half of this amount | £1 |

Add these amounts up, obtaining £7. It should be the same as your previous answer.
 Test this method on some other purchases. Does it always work? Can you explain why it works?

## Exercise ME

In this chapter we considered simple mortgages. We only considered the first few years of a mortgage – we did not find out when the mortgages would be paid off, nor the annual repayments needed to pay them off in a fixed time. A spreadsheet can be used for many of the calculations involved.

Suppose the mortgage is for £30 000, at a constant rate of 8%. For the moment, suppose the annual repayment is £3000.

Enter 30 000 in A1, and 3000 in B1. The amount outstanding at the end of one year is $1.08 \times £30\,000 - £3000$. So in A2 enter =1.08*A1-B$1.

Copy this down the A column to see the amount outstanding after each year. When is the loan paid off?

Adjust the annual repayment in B1 so that the loan is paid off in precisely 25 years.

What is the effect of adjusting the interest rate?

This is a very simple model of a mortgage. What factors have we ignored?

**You will need:**
- computer with spreadsheet package installed

# Review section

The chapters of this book have covered a lot of new mathematics. But for the GCSE exam you also need to know the mathematics that was covered in previous years. Some of this has already been revised, for example trigonometry was revised in chapter 2 on the sine and cosine rules. The following review modules complete the revision of topics that you need.

There are four review modules, covering Number, Algebra, Shape and space, and Handling data. Each module contains several topics. Each topic is organised as follows.

- A brief introduction about the topic
- Definitions, results and methods: details of the formulae and results you may need, and the techniques used in the topic
- Worked examples
- Exercise
- Question spotting: this tells you how to identify the topic in an exam paper, so that you can immediately employ the correct methods
- Tips: some hints to help you deal successfully with problems in the topic
- Links: indications of when the topic might be associated with others

The most important thing is for you to practise by doing the questions in the exercises. Some of these questions are to be done with a calculator, and some without. Some of the questions are split up into several parts to lead you through; in others you are expected to supply the links yourself.

# Review module 1 – Number

The topics covered in this module are:

- **fractions**
- **factors and multiples**
- **error and limits of accuracy**
- **approximation**
- **appropriate accuracy**
- **standard form**
- **proportion.**

## Fractions

Numbers less than 1 can be represented as decimals or as **fractions**. In this section we revise the arithmetic of fractions.

### Definitions and methods

- The top number of a fraction is the **numerator**.
- The bottom number of a fraction is the **denominator**.
- A number with both an integer and a fraction, such as $6\frac{1}{7}$, is a **mixed number**.
- A number in which the numerator is greater than the denominator, such as $\frac{4}{3}$, is an **improper fraction**.
- To multiply fractions, multiply the numerators and the denominators.
- To divide fractions, turn the second fraction upside down and then multiply.
- To add or subtract fractions, adjust so that they have the same denominator. Then add or subtract the numerators.

*Example*  Find **a** $4\frac{1}{5} \div 3\frac{3}{4}$  **b** $4\frac{1}{5} - 3\frac{3}{4}$

**a** First convert both mixed numbers to improper fractions.

$$4\frac{1}{5} = \frac{4 \times 5 + 1}{5} = \frac{21}{5}$$

$$3\frac{3}{4} = \frac{3 \times 4 + 3}{4} = \frac{15}{4}$$

Turn the second fraction upside down and multiply.

$$4\frac{1}{5} \div 3\frac{3}{4} = \frac{21}{5} \times \frac{4}{15}$$
$$= \frac{84}{75}$$
$$= \frac{28}{25}$$

$4\frac{1}{5} \div 3\frac{3}{4} = 1\frac{3}{25}.$

**b** Subtract the integer parts.

$$4 - 3 = 1$$

Subtract the fraction parts, converting them to a common denominator of 20.

$$\tfrac{1}{5} - \tfrac{3}{4} = \tfrac{4}{20} - \tfrac{15}{20}$$
$$= -\tfrac{11}{20}$$

Add this to 1, i.e. subtract $\tfrac{11}{20}$ from 1.
$4\tfrac{1}{5} - 3\tfrac{3}{4} = \tfrac{9}{20}$.

## Exercise on fractions

1. Evaluate the following.
   - **a** $\tfrac{7}{8} \times \tfrac{2}{3}$
   - **b** $\tfrac{3}{4} \div \tfrac{1}{7}$
   - **c** $\tfrac{5}{6} \div \tfrac{1}{5}$
   - **d** $\tfrac{1}{4} + \tfrac{1}{3}$
   - **e** $\tfrac{3}{10} + \tfrac{2}{7}$
   - **f** $\tfrac{5}{12} - \tfrac{1}{9}$
   - **g** $1\tfrac{1}{2} + 2\tfrac{3}{4}$
   - **h** $3\tfrac{2}{5} + 4\tfrac{7}{10}$
   - **i** $1\tfrac{2}{3} + 2\tfrac{2}{7}$
   - **j** $3\tfrac{1}{10} - 1\tfrac{1}{2}$
   - **k** $4\tfrac{1}{4} - 2\tfrac{1}{12}$
   - **l** $5\tfrac{1}{3} - 4\tfrac{7}{10}$
   - **m** $4\tfrac{1}{2} \times 2\tfrac{1}{3}$
   - **n** $1\tfrac{1}{10} \times 2\tfrac{1}{6}$
   - **o** $3\tfrac{3}{7} \times 2\tfrac{4}{5}$
   - **p** $2\tfrac{7}{9} \div 4\tfrac{1}{3}$
   - **q** $3\tfrac{2}{5} \div 4\tfrac{3}{8}$
   - **r** $4\tfrac{7}{8} \div 2\tfrac{3}{5}$

2. In a horse race, the winner Red Brandy finished $3\tfrac{1}{2}$ lengths ahead of the second horse Zohar, which finished $2\tfrac{2}{3}$ lengths ahead of the third horse Colleen. By how much did Red Brandy beat Colleen?

3. A jar contains $6\tfrac{5}{8}$ litres of juice. How many full glasses of $\tfrac{3}{5}$ litre each can be taken from the jar?

4. An exam lasts $2\tfrac{1}{4}$ hours, of which the first $\tfrac{1}{3}$ hour is for reading the questions before writing anything down. How much time is allowed for answering the questions?

5. A jar holds $3\tfrac{1}{5}$ ounces of flour. A quantity of flour is measured out by filling the jar $4\tfrac{1}{2}$ times. How much flour is obtained?

### Question spotting
Watch out for any question in which the quantities are given as fractions or as mixed numbers.

### Tips
When multiplying or dividing mixed numbers, first convert them to improper fractions. In any calculation with fractions, show all your working. Give your answer as a fraction, not as a decimal.

### Links
Manipulation of fractions is likely to be required as part of another question, involving distances, times, volumes, probabilities etc.

# Factors and multiples

If $n$ divides exactly into $m$, then $n$ is a factor of $m$, or equivalently $m$ is a multiple of $n$. Note that in this topic we consider positive whole numbers, not fractions or negative numbers.

## Definitions

- A **prime number** has only two factors, 1 and itself. (Note that 1 is *not* a prime number.)
- A **composite number** is a number with more than two factors.
- The **highest common factor** (HCF) of two or more numbers is the largest number which is a factor of all of them.
- The **least common multiple** (LCM) of two or more numbers is the least number which is a multiple of all of them.

## Results

Any number can be written as the product of primes. Below are tests to find simple prime factors.

A number is even (divisible by 2) if its last digit is 0, 2, 4, 6 or 8.
A number is divisible by 3 if the sum of its digits is divisible by 3.
A number is divisible by 5 if its last digit is 0 or 5.

To find the HCF or LCM of numbers, first express them as products of primes.

---

*Examples* Find the HCF and LCM of 72 and 300.

Factorise both the numbers. We know that 72 is divisible by 2 (as its last digit is 2) and by 3 (the sum of its digits is 9, which is divisible by 3). So keep on dividing by 2 and then keep on dividing by 3.

$$72 = 2 \times 36 = 2 \times 2 \times 18 = 2 \times 2 \times 2 \times 9 = 2 \times 2 \times 2 \times 3 \times 3 = 2^3 \times 3^2$$

Similarly, 300 is divisible by 2, 3 and 5. Keep on dividing by these numbers.

$$300 = 2 \times 150 = 2 \times 2 \times 75 = 2 \times 2 \times 3 \times 25 = 2 \times 2 \times 3 \times 5 \times 5 = 2^2 \times 3 \times 5^2$$

For the HCF, take the smallest power of each prime which is a factor of *both* numbers. The smallest power of 2 is $2^2$, and the smallest power of 3 is 3 itself.

$$2^2 \times 3 = 4 \times 3 = 12$$

The HCF is 12.
   For the LCM, take the largest power of each prime which is a factor of *either* number. So take $2^3$, $3^2$ and $5^2$.

$$2^3 \times 3^2 \times 5^2 = 8 \times 9 \times 25 = 1800$$

The LCM is 1800.

---

The European Snap championships are held every 14 years. The Commonwealth Tiddlywinks finals are held every 12 years. These events coincided in 2000 – when will they next be held in the same year?

The events are held every multiple of 14 years and 12 years respectively. So they will be held simultaneously after the least common multiple of 14 and 12.

$$14 = 2 \times 7$$
$$12 = 2^2 \times 3$$

The LCM is $2^2 \times 3 \times 7 = 84$.
The events will be held in the same year in 2084.

## Exercise on factors and multiples

1. Express each of the following numbers as products of powers of primes.
   **a** 20     **b** 44     **c** 70     **d** 60     **e** 156
2. Find the HCF of each of the following pairs of numbers.
   **a** 12 and 18   **b** 20 and 35   **c** 33 and 26   **d** 15 and 60
3. Find the LCM of each of the pairs of numbers in question 2.
4. Find the HCF and LCM of each of the following sets of numbers.
   **a** 6, 14, 20   **b** 15, 6, 8   **c** 4, 20, 14
5. A dial has two hands, which rotate through a full circle every 10 seconds and every 12 seconds respectively. The hands start together at the top of the dial. When will they next be together at the top of the dial?
6. Beads, of different colours and shapes, are threaded on a wire. Every sixth bead is blue, and every eighth bead is square. The first bead is blue and square. Which is the next blue, square bead?
7. Jean and Robin agree to meet for lunch every ninth day, provided that that day is a Saturday. Their first meeting is on Saturday 1 January. When is their next meeting?
8. A lighting system has red, blue and yellow lights which flash every 9 seconds, 15 seconds and 21 seconds respectively. Initially they all flash together. After how long will they next flash together?
9. A wall is 156 cm by 144 cm. It is to be covered by equal square tiles (without breaking any). What is the largest square that can be used?
10. At a children's party there are 18 girls and 24 boys. They are to be divided into teams of equal size, which are either all-girl or all-boy. What is the largest possible size of team, if no child is left out?
11. There are three pieces of string, of lengths 75 cm, 50 cm and 60 cm. They are to be cut up into segments of equal length. What is the greatest possible length of the segments, if no string is wasted?

### Question spotting

The phrases 'highest common factor' or 'least common multiple' may occur in a question. Otherwise, be on the lookout for a problem which asks you to find the smallest number that certain given numbers will divide into (LCM) or the largest number which will divide into the given numbers (HCF). The given numbers might involve time or distance.

### Tips

When finding prime factors, start with the simple factors of 2, 3 and 5. After removing all these factors, see if 7 or 11 is a factor. You won't be asked to factorise something like 9367 (which is $17 \times 19 \times 29$).

When finding the HCF, take the primes which occur in *all* the original numbers. When finding the LCM, take the primes which occur in *any* of the original numbers.

# Error and limits of accuracy

Any measurement is liable to error. Similarly, if the result of a calculation is rounded, then there is a possible error.

### Definitions

- The positive difference between the true value and the measured value is the **error** (or absolute error).

### Results

Suppose a quantity is measured to an accuracy of a certain value. Then the limits of the true quantity are half that value on either side of the measurement.

Suppose a weight is given as 3.7 kg, correct to one decimal place. Then the limits of the weight are 3.65 kg to 3.75 kg. Note that the weight could be 3.65 kg (which would be rounded up to 3.7 kg) but not 3.75 kg (which would be rounded up to 3.8 kg).

Suppose two quantities are added or subtracted. Then the maximum error is the sum of the individual maximum errors. (Note: do not subtract the errors.)

Suppose two quantities are multiplied. The limits of the product are found by multiplying the two maximum values, and multiplying the two minimum values.

Suppose one quantity is divided by another. The top limit is the maximum value of the first quantity divided by the *minimum* value of the second. The bottom limit is the minimum value of the first quantity divided by the *maximum* value of the second.

---

*Example*  A police speed trap measured the distance moved by a car as 21 m, to the nearest 1 m, over a period of 0.9 seconds, correct to one decimal place.
**a** Find the upper and lower limits on the speed of the car.
**b** Comment on a claim that the car was travelling at more than 22 m/s.

**a** The upper limit is found by taking the greatest distance, divided by the least time.

$$21.5 \div 0.85 = 25.3$$

The lower limit is found by taking the least distance, divided by the greatest time.

$$20.5 \div 0.95 = 21.6$$

The limits of the speed are 25.3 m/s and 21.6 m/s.
**b** Notice that 22 is greater than the lower limit.
It is not certain that the car was travelling at more than 22 m/s.

---

## Exercise on errors and limits of accuracy

**1** In the following, write down the limits between which the true value lies.
  **a** The crowd at a football match, given as 23 000 to the nearest 1000
  **b** The weight of a parcel, given as 43 grams to the nearest gram
  **c** A distance between towns, given as 35 miles to the nearest 5 miles
  **d** The time of a race, given as 12.4 seconds to one decimal place
  **e** A temperature, given as 32.7°C correct to one decimal place
  **f** The mass of a chemical, given as 0.23 grams correct to two decimal places
  **g** An electrical current, given as 0.0307 amps, correct to three significant figures.

2 A car has mass 1200 kg, to the nearest 50 kg, and the driver and passengers have mass 260 kg, to the nearest 10 kg. Find the limits between which the total mass lies.

3 A film has length 93 minutes, correct to the nearest minute. When it is shown on television it is interrupted by advertisements which last a total of 20 minutes, correct to the nearest 5 minutes. Find the limits between which the total time lies.

4 The distance between two walls of a bathroom is 230 cm. A rail for a shower curtain will be fitted between the walls. The rail is 229 cm long. Both measurements are given to the nearest centimetre. Will the rail fit between the walls? Give reasons.

5 A train will leave a station at 08.18 at the earliest. Deirdre leaves home at 07.30, to the nearest 5 minutes, and drives to the station, her journey taking 45 minutes, to the nearest 5 minutes. Is she certain to catch the train? Give reasons.

6 A book is expected to have 380 pages, correct to the nearest 10 pages, and each page will contain an average of 350 words, correct to the nearest 10 words. Find the limits between which the total number of words lies.

7 The maximum load of a lorry is 3 tonnes. It is to carry 1600 items, to the nearest 100, each of which has mass 1.85 kg, correct to two decimal places. Will the lorry be overloaded? Give reasons.

8 A manuscript has 76 000 words, correct to the nearest 1000. When it is made into a book, the average number of words per page will be 350, correct to the nearest 10. Will the book exceed 220 pages? Give reasons.

9 A file on the internet has length 27 000 kilobits, to the nearest 1000 kilobits. It is downloaded at a speed of 45 kps (kilobits per second), to the nearest 5 kps. Will the downloading be finished within 11 minutes? Give reasons.

10 At the beginning of the day, Jacob had £80, correct to the nearest £5. He went on an outing, spending £60, correct to the nearest £10. Will he have enough left for his £13.50 fare home?

## Question spotting

Words like 'accuracy' or 'limit' may not appear in a question. Watch out for a situation in which the numbers are given to a certain number of decimal places or significant figures, and which asks you whether a certain result is possible.

## Tips

Don't forget to halve the degree of accuracy when finding the maximum error. If a value is given as 3.7, to one decimal place, then the limits are 3.65 to 3.75, not 3.6 to 3.8.

Remember to *add* the errors when subtracting measurements.

For the upper limit of a ratio, remember to take the upper limit of the numerator divided by the *lower* limit of the denominator.

Don't forget to put in the *units* of an answer. For example, if the question involves mass, give the answer as 5 kg, not as 5.

Don't give your answer to a high degree of accuracy. In the example above, the lower limit of the speed was $20.5 \div 0.95$, which to six decimal places is 21.578947. This is far more precise than the situation justifies, and so we rounded to 21.6. (This is less than 22, which was significant for the second part of the question.)

## Links

The topic of accuracy is linked with any situation involving measurement, such as area and volume, rates of change and so on.

## Approximation

We can find the approximate value of a calculation by simplifying each number involved. This is useful when we want to check that the results given by a calculator are correct, or when we do not have a calculator available.

### Method

Round each number to one significant figure, and then do the calculation mentally or on paper.

*Example*  Pauline's bus fares to work are £2.35 per day. She goes to work five days per week. She calculates that over a period of 19 weeks she will spend £2232.50. Make a rough estimate of the amount to see whether her calculation is correct.

Round each number to one significant figure, then multiply together.

2.35, 5 and 19 are approximately 2, 5 and 20 respectively
$2 \times 5 \times 20 = 10 \times 20 = 200$

So the amount is approximately £200. Pauline is out by a factor of 10. (Perhaps she entered 23.5 instead of 2.35.)

## Exercise on approximation

1 Find the approximate value of each of the following:
   a $48.34 \times 82.1$
   b $(2.54 + 4.33)^2$
   c $\dfrac{38.4 \times 5.77}{0.362}$

2 Find the approximate number of seconds in a month.
3 Find the approximate cost of 67 textbooks at £8.99 each.
4 About how long will it take to download a file of 78 000 kilobits at a speed of 42 kilobits per second?
5 A distance of 2360 metres was run in 527 seconds. What was the approximate average speed?
6 About how long will it take to drive 267 km at an average speed of 83 km/h?
7 Mr McCluskey works for 216 hours at an hourly wage of £7.34. Find a rough value for the amount he earns.
8 In an income scheme there is a tax-free allowance of £4721, and income over that is taxed at 27%. Roughly how much would a man earning £26 321 expect to pay in tax?
9 Samantha goes on holiday to Ruritania, where the exchange rate is 3472 crowns to the £.
   a About how much will she get for £426?
   b At the end of her holiday she has 263 500 crowns left. About how much will she get if she changes them back to £ at the same rate?
10 The expression $(12.54 + 37.33)^2$ was evaluated, and the result was given correct to three significant figures. Which of the answers below is correct? Give your reasons.
   a 27.3   b 30 200   c 9320   d 2490
11 Which of the answers below is the value of $\sqrt{5.382 \times 7.843}$, correct to three significant figures? Give your reasons.
   a 7.32   b 6.50   c 63.1   d 5.31   e 58.4

### Question spotting

In any question about approximation there will appear a key phrase such as 'approximate', 'rough value', or 'about how much', etc.

### Tips

Do not expect too much of approximation. It checks that an answer is not 10 times too large or too small.

Be sure to give units in your answer. For example, if the question involves speed give the answer as 4 m/s, not as 4.

### Links

Questions concerning approximation could come in connection with many situations involving money, measurement and so on. Even if you are not asked to do an approximation, it is good practice to check that your answer is approximately correct.

## Appropriate accuracy

The result of any calculation should be given to a degree of accuracy which is justified by the context.

### Method

At the end of a calculation, give an answer to the degree of accuracy which you can guarantee to be correct. Suppose you give a length as 2.32 metres, correct to two decimal places. Then those digits should be correct – the length is not 2.31 metres, or 2.33 metres. If you think that the length could be anywhere between 2.31 and 2.34 metres, then give the answer as 2.3 metres, i.e. to an accuracy of one decimal place.

If the numbers in a calculation are given to a certain degree of accuracy, say to three significant figures, then the final result should not be to a greater degree of accuracy.

*Examples*  A car is driven for 290 km, at an average speed of 94 km/h. Find the time taken, giving your answer to an appropriate degree of accuracy.

The time taken is the distance divided by the speed. To six significant figures,

$$290 \div 94 = 3.08511$$

This answer is far too accurate. The original values were given to two significant figures, so give the answer to the same degree of accuracy.

The time taken was 3.1 hours.

**Note.** With this sort of question, there is often no single correct answer. Here it would also be correct to give the answer as 3 hours.

The exchange rate of pounds to euro is €0.62946 for £1. I have about £360 to change to euro. How much will I get?

Divide 360 by 0.62946.

$$360 \div 0.62946 = 571.919 \text{ to six significant figures}$$

Though the exchange rate is given accurately, the number of pounds is approximate. The accuracy of the result of a calculation is the accuracy of the *least* precise part of the calculation. In this case only two significant figures are justified.
I will get €570.

---

Make a rough estimate of $\dfrac{4.36 \times 32.6}{205}$. Then use your calculator to evaluate the expression, giving your answer to an appropriate degree of accuracy.

Round 4.36, 32.6 and 205 to 4, 30 and 200 respectively.

$$\frac{4 \times 30}{200} = \frac{12}{20} = 0.6$$

$\dfrac{4.36 \times 32.6}{205}$ is approximately 0.6.

Note that the numbers are given to three significant figures. So that is an appropriate degree of accuracy for the answer. Use your calculator to work out the expression, giving your answer correct to three significant figures.

$$\frac{4.36 \times 32.6}{205} = 0.693.$$

**Note.** This result is close to the approximate value of 0.6.

---

# Exercise on appropriate accuracy

In all these questions, give your answer to an appropriate degree of accuracy.

**1** Evaluate the following.
   **a** $3.21 \times 4.87$        **b** $62.45 \div 21.441$        **c** $4.32946 \times 420$
   **d** $3.29564 \div 700$        **e** $\dfrac{29 \times 47}{0.32 \times 0.72}$        **f** $\sqrt{3.8^2 + 2.7^2}$

**2** A cake of mass 1000 grams is divided into three equal portions. What is the mass of each portion?
**3** When a garden was dug up, the earth that was removed had mass 3.2 tonnes and volume $1.1\,\text{m}^3$. What was the density of the earth?

> **Remember:**
> density = mass ÷ volume

**4** In a week, a farm produced 1878 sacks of potatoes, weighing about 17 kg each. What was the total weight of the potatoes?
**5** Julian walked for 3 hours and 20 minutes, at a speed of 4 miles per hour. How far did he walk?
**6** The midday temperature was found every day for a week. The results, in °C, are below. What was the average temperature?

   23   21   24   26   23   25   22

**7** At a certain time there are ¥163.47 (Japanese yen) to £1.
   **a** Mr Sampson has about £740 to change to yen. How much will he get?
   **b** Miss Hamano changes about ¥450 000 to pounds. How much will she get?

8   A current of $I$ amperes through a circuit of resistance $R$ ohms will generate power of $RI^2$ watts. Find the power generated by 3.7 amps through a circuit of resistance 12 ohms.
9   The density of iron is 7874 kg/m$^3$.
    a   A lump of iron has volume about 0.2 m$^3$. What is its mass?
    b   An iron girder has mass about 1.4 tonnes. What is its volume?
10  How many days are there in an average lifetime of 70 years?
11  One mile is 1.6093 kilometres, and 1 kilogram is 2.2046 pounds.
    a   The distance from Belfast to Enniskillen is 70 miles. What is this in kilometres?
    b   The distance from Paris to Lyon is 400 km. What is this in miles?
    c   Michael weighs 128 pounds. What is that in kilograms?
    d   Astrid weighs 53 kg. What is that in pounds?

### Question spotting
You are unlikely to get a question which is purely about appropriate accuracy. The principle of giving an answer to an appropriate degree of accuracy holds throughout mathematics. It applies for any calculation in which the result is not exact.

### Tips
In all calculations think about the level of accuracy that is appropriate, and round the answer to a suitable number of decimal places or significant figures.

Be sure to give units in your answer. For example, if the question involves time give the answer as 5.3 seconds, not as 5.3.

Remember that the accuracy of the answer to a calculation should be that of the *least* accurate of the original numbers.

### Links
The principle of giving an answer to an appropriate degree of accuracy applies in connection with any question which asks you to find a length, a mass, an amount of money and so on.

## Standard form

Any number can be written in standard form. It is particularly useful for very large or very small numbers.

### Definition
- A number is in standard form if it consists of a number between 1 and 10, multiplied by a power of 10.

### Results
To convert a large number to standard form, move the decimal point to the left until there is only one digit to its left. The power of 10 is given by the number of moves.

$$234\,000\,000 = 2.34 \times 10^8$$

To convert a small number to standard form, move the decimal point to the right until there is only one digit to its left. The power of 10 is negative, and is given by the number of moves.

$$0.000000282 = 2.82 \times 10^{-7}$$

To multiply/divide two numbers in standard form, multiply/divide the number parts and add/subtract the indices.

$$2.3 \times 10^8 \times 2 \times 10^9 = 4.6 \times 10^{17} \qquad (6.3 \times 10^8) \div (3 \times 10^3) = 2.1 \times 10^5$$

After a multiplication or division, the result may not be in standard form. Adjust the number part and the power of 10.

$$6.3 \times 10^8 \times 2 \times 10^5 = 12.6 \times 10^{13} = 1.26 \times 10^{14}$$
$$(2.4 \times 10^{11}) \div (6 \times 10^3) = 0.4 \times 10^8 = 4 \times 10^7$$

To add/subtract two numbers in standard form, first adjust the number with the lower power of 10. Then add/subtract. For $6.4 \times 10^8 + 3.2 \times 10^7$, adjust the second term so that it involves the same power of 10 as the first term.

$$6.4 \times 10^8 + 3.2 \times 10^7 = 6.4 \times 10^8 + 0.32 \times 10^8 = 6.72 \times 10^8$$
$$5.3 \times 10^9 - 7 \times 10^8 = 5.3 \times 10^9 - 0.7 \times 10^9 = 4.6 \times 10^9$$

The examples above all involve large numbers in standard form, i.e. those with a positive index. The same results are true for small numbers, i.e. those with a negative index. But be careful with the arithmetic of negative numbers.

$$\times \quad 2 \times 10^{-5} \times 3 \times 10^{-7} = 6 \times 10^{-12} \quad (-5 + -7 = -12)$$
$$\div \quad (5 \times 10^{-3}) \div (2 \times 10^{-11}) = 2.5 \times 10^8 \quad (-3 - (-11) = -3 + 11 = 8)$$

Adjustment $\quad 5 \times 10^{-6} \times 3 \times 10^{-3} = 15 \times 10^{-9}$
$$= 1.5 \times 10^{-8} \quad (-9 + 1 = -8)$$
$$(2 \times 10^{-17}) \div (5 \times 10^{-4}) = 0.4 \times 10^{-13}$$
$$= 4 \times 10^{-14} \quad (-13 - 1 = -14)$$

Adding/subtracting $\quad 3.2 \times 10^{-11} + 5.7 \times 10^{-12} = 3.2 \times 10^{-11} + 0.57 \times 10^{-11}$
$$= 3.77 \times 10^{-11}$$

*Examples*   A star is $2.4 \times 10^{16}$ km away. A spaceship can travel at 600000 km per second. How long will it take the spaceship to reach the star? Give your answer in standard form.

To find the time taken, divide the distance by the speed. First put the speed into standard form.

$$600000 = 6 \times 10^5$$
$$(2.4 \times 10^{16}) \div (6 \times 10^5) = 0.4 \times 10^{11}$$
$$= 4 \times 10^{10}$$

**Remember:**
time = distance ÷ speed

The spaceship will take $4 \times 10^{10}$ seconds.

---

The mass of a ship, when empty, is $2.4 \times 10^7$ kg. The mass of its cargo is $7.3 \times 10^6$ kg. Find the mass of the laden ship.

Add these two numbers. Adjust the second number so that it has the same power of 10 as the first number.

$$2.4 \times 10^7 + 7.3 \times 10^6 = 2.4 \times 10^7 + 0.73 \times 10^7$$
$$= 3.13 \times 10^7$$

The total mass is $3.13 \times 10^7$ kg.

# Standard form

## Exercise on standard form

1 Evaluate these expressions, leaving your answers in standard form.
   a $4.2 \times 10^9 \times 2 \times 10^8$
   b $2.4 \times 10^5 \times 3 \times 10^5$
   c $5.3 \times 10^8 \times 2 \times 10^6$
   d $8 \times 10^6 \times 7 \times 10^{11}$
   e $(6.6 \times 10^7) \div (3 \times 10^5)$
   f $(8.2 \times 10^{11}) \div (2 \times 10^3)$
   g $(3 \times 10^5) \div (4 \times 10^2)$
   h $(2 \times 10^{13}) \div (5 \times 10^4)$
   i $3 \times 10^6 + 4 \times 10^6$
   j $8 \times 10^7 + 6 \times 10^7$
   k $4.1 \times 10^8 + 5 \times 10^7$
   l $7.2 \times 10^{11} - 9 \times 10^{10}$
   m $4.9 \times 10^6 + 2 \times 10^5$
   n $3.2 \times 10^{-5} \times 2 \times 10^{-9}$
   o $3.5 \times 10^8 \times 2 \times 10^{-21}$
   p $(4.5 \times 10^4) \div (3 \times 10^{-8})$
   q $(2.8 \times 10^{-3}) \div (4 \times 10^{-5})$
   r $3 \times 10^{-7} + 2 \times 10^{-7}$
   s $6 \times 10^{-5} + 8 \times 10^{-5}$
   t $6.8 \times 10^{-7} - 2 \times 10^{-8}$
   u $4 \times 10^{-11} - 2 \times 10^{-10}$

2 A spaceship can travel at 1 200 000 m/s. How far will it travel in $3 \times 10^{13}$ seconds?

3 How long will it take the spaceship of question 2 to travel $4.8 \times 10^{23}$ m?

4 Let $K = \dfrac{a \times b}{c}$. Evaluate $K$ in each of the following cases, leaving your answers in standard form.
   a $a = 4 \times 10^{11}, b = 3 \times 10^4, c = 6 \times 10^8$
   b $a = 5 \times 10^{16}, b = 2 \times 10^{11}, c = 4 \times 10^5$
   c $a = 3 \times 10^{-3}, b = 6 \times 10^{-7}, c = 9 \times 10^7$
   d $a = 8 \times 10^{-4}, b = 6 \times 10^{-3}, c = 4 \times 10^{11}$

5 The mass of an oxygen atom is about $2.5 \times 10^{-26}$ kg.
   a What is the mass of $6 \times 10^{14}$ oxygen atoms?
   b How many atoms are there in $2 \times 10^7$ kg of oxygen?

Where relevant, give your answer correct to three significant figures.

6 Evaluate the following, leaving your answers in standard form.
   a $1.245 \times 10^{83} \times 8.632 \times 10^{32}$
   b $(5.331 \times 10^{121}) \div (8.112 \times 10^{-17})$
   c $1.22 \times 10^{-77} \times 3.03 \times 10^{-41}$
   d $(5.441 \times 10^{-81}) \div (2.07 \times 10^{34})$

7 How many seconds are there in 15 billion years?

8 The distance between Coleraine and Dungannon is 75 km. How long would it take a snail to slither this distance at $4.2 \times 10^{-2}$ m/s?

9 The radius of the Sun is about $6.96 \times 10^8$ m. Find its volume.

### Question spotting

There should be no difficulty in spotting a question about standard form. Either the phrase standard form will occur, or the numbers involved will be given in standard form. Sometimes it is referred to as **standard index form**.

### Tips

Read the question carefully. In particular, if you are asked to give the answer in standard form, then do not give it as an ordinary number. Do not write 4 300 000, for example. Neither should you write $43 \times 10^5$. Convert to $4.3 \times 10^6$.

Your calculator may display numbers as, for example, $4.3^{06}$. Do not give this as the answer. Write it out properly, as $4.3 \times 10^6$.

Be careful with negative numbers.

Make sure you do the correct operation. If you find it hard to think of very large numbers, try smaller, more familiar ones. In the example above about stellar travel, it is difficult to imagine these immense distances and speeds. Think of a more homely journey – a distance of 100 miles at 50 m.p.h. will take 2 hours. To find the time, *divide* the distance by the speed.

## Links

Standard form can be involved in any problem in which there are very large or very small numbers. So you could find the topic occurring in connection with

- area and volume
- distance, speed, time.

# Proportion

Proportion connects the rate of change of two variables. There are different sorts of proportion.

## Definitions

- Two quantities are **directly proportional** if their ratio is constant. So if two quantities are directly proportional they are increasing together, so that if one is doubled the other is doubled. If $y$ and $x$ are directly proportional, we write $y \propto x$. This proportionality statement is not an equation. It is equivalent to the equation $y = kx$, where $k$ is the **constant of proportionality**. The constant $k$ can be found from a pair of values of $x$ and $y$.
- Two quantities are **inversely proportional** if their product is constant. So if two quantities are inversely proportional they increase in opposite directions, so that if one is doubled the other is halved. We write $y \propto \frac{1}{x}$, which is equivalent to $y = \frac{k}{x}$.
- In some cases one quantity is proportional to a power of another. For example

    if $y$ is proportional to the square of $x$, then $y \propto x^2$, or $y = kx^2$

    if $y$ is inversely proportional to the cube of $x$, then $y \propto \frac{1}{x^3}$, or $y = \frac{k}{x^3}$

---

*Examples*  I change £360 into Ruritanian crowns, and get 840 000. At the same exchange rate, how much would I need to change to get 1 200 000 crowns?

The number of crowns is proportional to the number of pounds. The constant of proportionality is the exchange rate, which is 840 000 crowns per £360. Simplify this exchange rate

$$840\,000 \div 360 = 2333\tfrac{1}{3} \text{ crowns per £}$$

Divide 1 200 000 by this exchange rate

$$1\,200\,000 \div 2333\tfrac{1}{3} = 514.29$$

I need to change £514.29.

A car with fuel economy 32 miles per gallon uses 1.5 gallons for a certain journey. What is the fuel economy of a car which does the same journey using 2 gallons?

The fuel economy of a car and the amount of fuel used are inversely proportional. The constant of proportionality is the product of the two.

$$32 \times 1.5 = 48$$

Divide 48 by 2, obtaining 24.
The fuel economy of the second car is 24 miles per gallon.

---

The mass of a solid metal cube is proportional to the cube of its side. A cube of side 5 cm has mass 1.3 kg.

**a** Find an equation giving the mass $m$ kg in terms of the side $s$ cm.
**b** What is the mass of a cube of side 4 cm?
**c** What is the side of a cube with mass 3.8 kg?

**a** We are told that $m \propto s^3$, or $m = ks^3$. Put $m = 1.3$ and $s = 5$.

$$1.3 = k5^3 \quad \text{so } k = 0.0104$$

The equation is $m = 0.0104 s^3$.

**b** Now put $s = 4$.

$$m = 0.0104 \times 4^3$$

The mass is 0.6656 kg.

**c** Now put $m = 3.8$.

$$3.8 = 0.0104 s^3 \quad \text{so } s^3 = \frac{3.8}{0.0104}, \text{ giving } s = \sqrt[3]{\frac{3.8}{0.0104}}$$

The side is 7.15 cm.

# Exercise on proportion

1 The electrical resistance $R$ ohms of a piece of wire is proportional to its length $l$ m. A piece of length 10 m has resistance 25 ohms.
  **a** Find $R$ in terms of $l$.
  **b** Find the length of wire for a resistance of 40 ohms.
2 $y$ is proportional to $x$. When $x = 7$, $y = 5$. Find an equation giving $y$ in terms of $x$.
3 The quantities $T$ and $S$ are inversely proportional. When $S = 5$, $T = 4$.
  **a** Find an equation giving $T$ in terms of $S$.
  **b** Find the value of $S$ when $T = 40$.
4 A certain journey takes 2.5 hours at 40 m.p.h. How long will the same journey take at 50 m.p.h.?

Where relevant, give answers correct to three decimal places.

5. For £438 I obtained 87 320 Japanese yen. Assuming that the money changer is not taking any commission, how many yen would I get for £391?

6. The volume $V\,\text{m}^3$ of a fixed mass of gas is inversely proportional to its pressure $P\,\text{N/m}^2$. At a pressure of $4\,\text{N/m}^2$ the volume is $2.7\,\text{m}^3$.
   a Find $V$ in terms of $P$.
   b Find the volume when the pressure is $1.2\,\text{N/m}^2$.

7. $y$ is proportional to the cube of $x$. When $x = 3$, $y = 54$.
   a Find $y$ in terms of $x$.
   b Find $y$ when $x = 0.8$.
   c Find the value of $x$ when $y = 2000$.

8. Discs are cut from a sheet of metal. The mass $m\,\text{g}$ of a disc is proportional to the square of the radius $r\,\text{cm}$. A disc of radius $2\,\text{cm}$ has mass $5\,\text{g}$.
   a Find $m$ in terms of $r$.
   b Find the mass of a disc of radius $7\,\text{cm}$.
   c Find the radius of a disc with mass $0.02\,\text{g}$.

9. $y$ is inversely proportional to the cube of $x$. When $x = 5$, $y = 4$. Find the value of $x$ for which $y = 3$.

10. The resistance $R$ ohms of a fixed length of wire is inversely proportional to the square of the radius $r\,\text{mm}$ of cross-section. When the radius is $0.4\,\text{mm}$ the resistance is $30$ ohms. What is the resistance if the radius is $0.15\,\text{mm}$?

11. The attraction $F\,\text{N}$ between two electrically charged particles varies inversely as the square of the distance $d$ metres between them. At a distance of $0.04\,\text{m}$ the attraction is $0.25\,\text{N}$. Find the distance at which the attraction is $25\,\text{N}$.

12. 'I see there,' said my father, 'a mine of problems for middle school:
    "A hunter, who possesses 720 grams of lead, can make 24 bullets for his gun. Knowing that the weight of the old-fashioned pound is 480 grams and that in this case, the number representing the bore of the gun also represents the number of bullets which one can make for the weapon with one pound of lead, what is the bore of the gun?"'
    *This worried me, as I feared that these problems would be tried out at the expense of my free time*
    Adapted from *La gloire de mon père*, by Marcel Pagnol.
    Find the solution to this problem.

### Question spotting

In situations involving simple or inverse proportionality, those words themselves may not appear. See the first two worked examples above. Be on the lookout for situations in which the variables change at the same rate or at opposite rates. In situations of proportionality to a power, the word 'proportion' will probably appear.

### Tips

You may prefer to write down a statement of proportion, and then find the constant of proportionality, or you may prefer to argue by common sense.

In this topic remember to provide units for answers, and to give answers to an appropriate degree of accuracy.

### Links

A question about proportion can occur in many contexts. In particular, area and volume problems often involve proportion to a square or a cube. See the review topic on similarity (page 294).

# Review module 2 – Algebra

The topics covered in this module are:

- **equations in one unknown**
- **simultaneous equations**
- **trial and improvement**
- **changing the subject**
- **inequalities**
- **linear programming**
- **graphs.**

## Equations in one unknown

When you solve an equation, you find the value or values of the unknown. Many real-life problems give rise to equations.

### Methods

When solving equations, the basic rule to follow is: *do the same thing to both sides.*

- If you add 2 to the right, you must also add 2 to the left.
- If you multiply the left by $(x + 3)$, you must multiply the right also by $(x + 3)$.
- If you square the left, you must also square the right.

Sometimes you need to use algebraic techniques like those covered in chapter 1.

- Expanding: for example, $3(4x - 2) = 12x - 6$.
- Collecting like terms: if you have $4x + 3x$, simplify to $7x$.
- Simplifying fractions: multiply both sides by the denominator of any fraction. This denominator could be either numerical or algebraic. For example

$$\frac{x+3}{2} = 5 \text{ becomes } x + 3 = 10, \qquad \frac{3}{x+1} = 4 \text{ becomes } 3 = 4(x+1)$$

---

*Examples*   Solve the following equations.

**a** $3x - 4 = 11$
**b** $2x - 1 = 14 - 3x$
**c** $2(x + 5) + 5(x - 2) = 21$
**d** $\dfrac{15}{x+1} = \dfrac{9}{x-1}$

**a** Add 4 to both sides, then divide by 3.

$$3x = 11 + 4 = 15$$

$x = 5$.

**b** It is better to collect the unknown letter on the side that gives it a positive coefficient. So collect the $x$s on the left, and the numbers on the right.

$$2x + 3x = 14 + 1$$
$$5x = 15$$

$x = 3$.

**c** Expand the brackets, then collect the terms.
$$2x + 10 + 5x - 10 = 21$$
$$7x = 21$$
$x = 3.$

**d** To clear up the fractions, multiply across by $(x - 1)$ and by $(x + 1)$.
$$15(x - 1) = 9(x + 1)$$
$$15x - 15 = 9x + 9$$
$$6x = 24$$
$x = 4.$

Check: put $x = 4$ into the original equation.
$$\text{Left side} = \frac{15}{4+1} = \frac{15}{5} = 3$$
$$\text{Right side} = \frac{9}{4-1} = \frac{9}{3} = 3$$

So the result is correct.

Lauren has three times as much money as Beatrice. If Lauren gives £20 to Beatrice, she will then have twice as much as Beatrice. If Beatrice has £$x$, find an equation in $x$. How much money does Lauren have?

If Beatrice has £$x$, then Lauren has £$3x$.
After Lauren has given £20 to Beatrice, Lauren has £$(3x - 20)$ and Beatrice has £$(x + 20)$. Lauren's amount is twice Beatrice's.
$$3x - 20 = 2(x + 20)$$

Now solve this equation.
$$3x - 30 = 2x + 40$$
$$x = 70$$

Now multiply this by 3.
Lauren has £210.

## Exercise on equations in one unknown

**1** Solve these equations.
  **a** $5x - 3 = 22$
  **b** $2x - 7 = 11$
  **c** $7 - 2x = 1$
  **d** $22 = 1 - 3x$
  **e** $3(4 + x) + 2(7 - x) = 43$
  **f** $2(3x - 1) - 5(2x + 3) = 11$
  **g** $\frac{1}{5}x - 2 = \frac{1}{6}x - 4$
  **h** $1 - \frac{2}{3}(1 - 5x) = 37$
  **i** $\frac{1}{4}(x - 6) + \frac{1}{2}(x + 3) = 9$
  **j** $\frac{1 - 2x}{8 - x} = 3$
  **k** $\frac{4 - 3x}{1 + x} = \frac{-7 - 3x}{7 + x}$
  **l** $\frac{4x - 3}{x} - \frac{2x - 1}{3x} = -2$

**2** The length of a rectangle is 3 m greater than the width. The perimeter is 54 m. If the width is $x$ m, find an equation in $x$ and solve it.

**3** In a general knowledge quiz, 3 marks are awarded for a correct answer, and 1 mark is deducted for an incorrect answer. Declan got twice as many answers right as he got wrong, and his final score was 85. How many questions did he answer in total?

**4** Aileen is 3 years younger than Brenda. Collette is 5 years younger than the total of the ages of Aileen and Brenda. The total of the ages of all the girls is 41. Find the ages of the girls.

**5** Four people, A, B, C and D, contribute money to their firm. The ratio of A's and B's contributions is 1:2, and the ratio of C's and D's contributions is 2:3. C contributes £7000 more than A. If the total amount is £45 000 find how much each person has contributed.

**6** There are 3 taxi firms in a town. Firm B has twice as many cars as Firm A. Firm C has 2 fewer cars than the total number of cars of Firms A and B. There are 34 cars in total. Find how many cars each firm has.

Where relevant, give your answers to three significant figures.

**7** Solve these equations.
  **a** $3x + 5 = 10x - 3$
  **b** $0.34x = 1.22x - 6.19$
  **c** $0.052(x - 3) = 1.332(x + 5)$
  **d** $\dfrac{x+4}{4.82} = \dfrac{7-x}{2.44}$
  **e** $\dfrac{3x - 1.441}{2x + 4.833} = 6$

**8** There are two parcels, A and B. Their total weight is 193 kg. Parcel A weighs 17 kg less than Parcel B. Find the weight of each parcel.

**9** A pile of 12 coins has total value £5.80. There are twice as many 20p coins as £1 coins, and the rest of the coins are all 50p coins. Find how many £1 coins there are.

**10** A library has CDs available for loan. Three fifths of them are classical, and a quarter of the rest are jazz. There are 42 CDs which are neither classical nor jazz. Find how many CDs there are in total.

**11** Allen went on a journey by walking, by train and by taxi. He walked to the station, which was $\frac{1}{15}$ of the total distance. Then he took a train, and at the station at the other end he took a taxi, which was $\frac{1}{18}$ of the total distance. The distance travelled by taxi was 0.3 miles less than the distance he walked. Find the total distance.

### Question spotting

There are several sorts of equation you could be asked to solve. There are the ones of this section, in which there is one solution for one unknown, there are the simultaneous equations of the next section, in which there are two unknowns, and chapter 3 of this book deals with quadratic equations which can have up to two solutions. Make sure you spot which sort of equation it is, and apply the appropriate techniques.

Some equation questions just ask you to solve an equation. Others are 'word problems' which require you to translate a situation described in words into an equation. Sometimes you will be told to take $x$ or another letter to stand for the unknown, in other cases you will have to provide that step for yourself.

In a 'word problem', provide the relevant units. If the answer to an equation is not exact, give it to an appropriate degree of accuracy.

### Links

An equation might be part of a question involving other topics in algebra. A word problem might arise in connection with area or volume, money matters and so on.

# REVIEW MODULE 2 – ALGEBRA

## Simultaneous equations

The equations of the previous section had only one variable. Simultaneous equations have two (or more) variables.

### Method

The usual technique to solve these equations is to eliminate one of the variables. You are then left with one equation in one unknown, which you solve by the methods of the previous section. Having found one variable, substitute it into one of the original equations to find the other variable.

You may be able to eliminate one of the variables immediately. You may have to multiply one or both of the equations to 'match' one of the variables so that it can be eliminated. If the variable you want to eliminate has different signs in the equations, then add the equations. If it has the same sign in both equations, then subtract the equations.

*Examples* Solve the equations

$$3x + y = 11 \quad [1]$$
$$2x - y = 4 \quad [2]$$

There is a single $y$ in both equations. Because $y$ is positive in the first equation and negative in the second, *add* the equations to eliminate $y$.

$$5x = 15 \quad [1] + [2]$$
$$x = 3$$

Substitute $x = 3$ in either of the equations. Substituting in the first equation

$$3 \times 3 + y = 11$$
$$y = 2$$

Hence $x = 3$ and $y = 2$.
Check: we used [1] to find $y$. Substitute the answers in [2] to check the result.

$$2 \times 3 - 2 = 4$$

This is correct.

Solve the equations

$$3x - 2y = 19 \quad [1]$$
$$4x - 5y = 30 \quad [2]$$

We cannot immediately eliminate either of the variables. If we seek to eliminate $y$, match the terms by multiplying the first equation by 5 and the second equation by 2.

$$15x - 10y = 95 \quad [3] = 5 \times [1]$$
$$8x - 10y = 60 \quad [4] = 2 \times [2]$$

Because the $y$-terms are both negative, i.e. they have the same sign, *subtract* the equations. Subtracting the second from the first, we have

$$7x = 35, \text{ giving } x = 5 \quad [5] = [3] - [4]$$

Substitute $x = 5$ into [1].

$$3 \times 5 - 2y = 19$$
$$-2y = 4, \text{ giving } y = -2$$

Hence $x = 5$ and $y = -2$.
Check: substitute in [2].

$$4 \times 5 - 5 \times (-2) = 20 + 10 = 30$$

This is correct.

---

For a concert, 800 tickets were sold, costing either £5 or £8. The total receipts were £4957. If $x$ of the cheaper tickets and $y$ of the more expensive tickets were sold, find equations in $x$ and $y$. Solve to find the number of more expensive tickets sold.

We have two unknowns, and two facts with which to find them. Using the fact that 800 tickets were sold, we have

$$x + y = 800 \qquad [1]$$

The revenue from the cheaper tickets was £$5x$, and from the more expensive tickets £$8y$. Using the fact that the total revenue was £4957, we have

$$5x + 8y = 4957 \qquad [2]$$

We are asked to find $y$. So it is more efficient to eliminate $x$. Multiply [1] by 5.

$$5x + 5y = 4000 \qquad [3]$$

Subtract [3] from [2].

$$3y = 957, \text{ giving } y = 319$$

319 of the more expensive tickets were sold.

## Exercise on simultaneous equations

**1** Solve these equations.
- **a** $3x + y = 7$
  $2x - y = 3$
- **b** $x + 3y = 13$
  $x + 5y = 17$
- **c** $4x + 3y = 15$
  $5x - 2y = 13$
- **d** $4x - 3y = 11$
  $3x + 2y = 4$
- **e** $3x - 5y = 1$
  $5x - 8y = 3$
- **f** $\frac{1}{2}x + \frac{1}{3}y = 9$
  $\frac{1}{4}x + \frac{1}{5}y = 5$
- **g** $0.1x - 0.3y = 2$
  $0.4x + 0.7y = 46$

**2** Three oranges and two lemons cost 81p. Two oranges and three lemons cost 79p. What is the cost of one orange?

**3** One shirt and two ties cost £30. Two shirts and three ties cost £53. Find the cost of a shirt. Find the cost of three shirts and four ties.

**4** In a tennis league 3 points are given for a win and 1 for a loss. (A tennis game cannot end in a draw.) After playing 20 matches, Jill has 42 points. How many games did she win?

**5** In a general knowledge test, 2 marks are given for a correct answer, and 1 mark is *deducted* for an incorrect answer. After answering 60 questions, Angela has 63 marks. How many did she get correct?

6. Kathleen has a pile of 215 coins, which are either 50p or 20p coins. The total value is £75.10. How many does she have of each?
7. In a football league, 3 points are given for a win, 1 for a draw and 0 for a loss. After playing 38 games, a team lost 11 games and gained 59 points. How many games did the team win?
8. Tickets for a cinema are either £5.40 or £6.70. A total of 426 tickets are sold, at a total cost of £2455.10. Find how many tickets were sold at each price.
9. The total weight of 6 similar cars and 7 similar vans is 19 110 kg. The weight of three of the cars is 1590 kg greater than the weight of one van. Find the weight of each vehicle.
10. The regulatory body of inter-city coach services imposes a fine of £247 for each late coach, and a reward of £13 for each coach which is on time. A coach firm makes 417 journeys, and makes a net payment of £559 to the regulatory body. How many coaches were late?

### Question spotting

You may be given a pair of simultaneous equations to solve. You may be given a 'word problem' like the one above about tickets. Be on the lookout for any situation in which there are two unknown quantities, and two facts about them.

### Tips

Before eliminating a variable, you must match up the coefficients.

If the variable you are going to eliminate has the same sign in both equations, then subtract the equations. If the signs are different, then add.

### Links

A question on simultaneous equations could be part of a general algebraic question. A word problem could involve money, like the one above, or lengths, masses, etc.

## Trial and improvement

This is a method of solving equations which cannot be solved exactly. By trying different values of $x$, and constantly improving the value, we can find a solution to any required degree of accuracy.

### Method

The method relies on the following.

> Suppose we want to solve $f(x) = k$. If $f(a) < k$ and $f(b) > k$, then there is a solution between $a$ and $b$.

Suppose you are asked to solve an equation to an accuracy of one decimal place. Try integer values of $x$, until you find a pair between which $f(x) - k$ changes sign. (In other words, $f(n) < k$ and $f(n+1) > k$, or vice versa.) Then there is a solution between $n$ and $n+1$. You might find, for example, that $f(3) < k$ but $f(4) > k$. So there is a solution between 3 and 4.

Now try values of $x$ given to one decimal place. As above, try until you have a pair of values between which $f(x) - k$ changes sign. Suppose $f(3.6) < k$ but $f(3.7) > k$. Then there is a solution between 3.6 and 3.7.

We have not finished. We want to know whether the solution is closer to 3.6 or 3.7. Test half-way, at 3.65. If $f(3.65) - k$ has the same sign as $f(3.6) - k$, then the solution lies between 3.65 and 3.7. To one decimal place, the solution is 3.7.

# Trial and improvement

*Example*  Solve the equation $x^3 - x = 1.6$, correct to one decimal place.

Fill in a table as below.

| x | $x^3 - x$ | conclusion |
|---|---|---|
| 1 | 0 | low |
| 2 | 6 | high |

So we know that the solution is between 1 and 2.

| 1.5 | 1.875 | high |
|---|---|---|
| 1.4 | 1.344 | low |

So the solution is between 1.4 and 1.5. Test half-way.

| 1.45 | 1.598625 | low |
|---|---|---|

> In fact, to 5 decimal places the solution is 1.45026.

So the solution is between 1.45 and 1.5.
To one decimal place, the solution is 1.5.

# Exercise on trial and improvement

Solve equations 1–4 correct to one decimal place.

**1** $x^3 + 3x = 20$
**2** $x^3 - 5x = 10$
**3** $2x^3 + x^2 = 8$
**4** $2x^3 - x = 43$

Solve equations 5–7 correct to two decimal places.

**5** $x^3 + x = 0.2$
**6** $2x^3 - x = 0.5$
**7** $10^x = 3$

> You will find a $10^x$ button on your calculator.

## Question spotting

The phrase 'trial and improvement' should appear in the question. You may be given a table to fill in.

## Tips

Do not forget the last step, of testing between 3.6 and 3.7. It is not enough to see which gives a closer value to $k$. In the example above, the value of $x^3 - x$ is closer to 1.6 when $x$ is 1.4 than when $x$ is 1.5, but the actual solution to the equation is closer to 1.5 than to 1.4. Hence the solution is 1.5, correct to one decimal place.

It is the value of $x$ which is to be given to one decimal place, not the value of f($x$). So keep on trying until you know the value of $x$ correct to one decimal place. Don't stop when your value of f($x$) is $k$, to one decimal place. In the example above, f(1.45) = 1.6, correct to one decimal place. But the answer is 1.5, not 1.45.

## Links

This topic could occur at the end of another question which asked you to set up an equation. For example, a problem in area or volume might give rise to an equation, which you would then be asked to solve using trial and improvement.

## Changing the subject

The subject of an equation is the letter expressed in terms of the other letters or numbers. When we change the subject, a different letter becomes the subject. This operation is also called **transposing** the equation.

### Method

The steps for changing the subject are exactly the same as for solving an equation. See the equations review topic (page 245). We obtain a new formula, instead of an actual number. If the new subject occurs more than once (such as $x$ in $ax + b = cx + d$), then collect the occurrences on one side and factorise.

*Examples*  Make $x$ the subject of the following equations.

**a** $y = ax - c$   **b** $y = \sqrt{z^2 - x^2}$   **c** $ax + b = cx + d$

**a** Add $c$ to both sides and divide by $a$.
$$x = \frac{y + c}{a}.$$

**b** Square both sides, get $x^2$ by itself on one side, then take the square root.
$$y^2 = z^2 - x^2$$
$$x^2 = z^2 - y^2$$
$$x = \sqrt{z^2 - y^2}.$$

**c** Collect the $x$-terms on the left and the other terms on the right. Factorise, then divide.
$$ax - cx = d - b$$
$$x(a - c) = d - b$$
$$x = \frac{d - b}{a - c}.$$

---

To hire a van for a day costs a fixed payment of £20 with £0.2 for each mile driven. Let the cost be £$C$ if the van was driven for $m$ miles.
**a** Find a formula giving $C$ in terms of $m$.
**b** Make $m$ the subject of this formula.

**a** To find the cost $C$, add the fixed payment to mileage payment multiplied by the number of miles.
$$C = 20 + 0.2m.$$
**b** Subtract 20 and divide by 0.2.
$$m = \frac{C - 20}{0.2} = 5(C - 20)$$
$$m = 5C - 100.$$

## Exercise on changing the subject

In questions 1–18, change the subject to the letter in brackets.

1. $y = 5x + k$    $(x)$
2. $t = am + 3$    $(m)$
3. $p = 3 - 2q$    $(q)$
4. $n = a - bm$    $(b)$
5. $z = \frac{1}{5}x + 3$    $(x)$
6. $v = \frac{2}{3}u - 8$    $(u)$
7. $y = 2\sqrt{x}$    $(x)$
8. $m = \sqrt{3k}$    $(k)$
9. $y = \sqrt{x} + 5$    $(x)$
10. $y = \sqrt{x - 5}$    $(x)$
11. $z = 3x^2$    $(x)$
12. $A = \pi r^2$    $(r)$
13. $a = \sqrt{b^2 + 4}$    $(b)$
14. $m = \sqrt{7 - n^2}$    $(n)$
15. $3x + 1 = 4 - ax$    $(x)$
16. $km - n = lm + gn$    $(m)$
17. $y = \dfrac{x + 5}{x - 7}$    $(x)$
18. $w = \dfrac{3z + a}{2z - b}$    $(z)$

19. A recipe for pork tells the cook to roast it for 20 minutes plus 30 minutes per pound. Suppose a joint of weight $p$ pounds is roasted for $m$ minutes.
    a Write down a formula giving $m$ in terms of $p$.
    b Make $p$ the subject of this formula.

20. A drainage repair service has a call-out fee of £25, then charges at a rate of £35 per hour. Suppose it charges £$C$ for $h$ hours work.
    a Write down a formula giving $C$ in terms of $h$.
    b Make $h$ the subject of this formula.

21. A simple income tax scheme levies tax at 25% of income over £4000. Let the income be £$I$ (where $I \geq 4000$) and the tax be £$T$.
    a Write down a formula giving $T$ in terms of $I$.
    b Make $I$ the subject of this formula.

22. At a restaurant there is a cover charge of £1.50. There is also a service charge, of 10% of the total cost of the meal itself and the cover charge. Suppose that the meal itself costs £$m$, and that the final cost including the service charge is £$C$.
    a Write down a formula giving $C$ in terms of $m$.
    b Make $m$ the subject of this formula.

23. The volume $V$ of an oblate spheroid (a sort of flattened sphere) is found by squaring the longest diameter $a$, multiplying this by the shortest diameter $b$, then multiplying by $\pi$ and dividing by 6.
    a Write down the formula giving $V$ in terms of $a$ and $b$.
    b Make $a$ the subject of this formula.

24. Suppose a weight is hung on the end of a string of length $l$ m. When it is allowed to swing, the period of swing, $T$ seconds, is found by dividing $l$ by 9.8, taking the square root of the result, and then multiplying by $2\pi$.
    a Write down a formula giving $T$ in terms of $L$.
    b Make $l$ the subject of this formula.

### Question spotting

A question on this topic might be in a practical context, as in the example above. Changing the subject can also be described as 'rearranging an equation', or 'transposing a formula', so watch out for any of these phrases.

### Tips

All the hints about solving equations apply also to changing the subject. In particular, make sure to do the same thing to both sides.

Be sure to include brackets where necessary. For example, when multiplying $y + b$ by $a$, the result is $a(y + b)$ or $ay + ab$, not $ay + b$.

When the new subject occurs more than once, you must collect together all the occurrences. If $x$ is expressed in terms of itself, as for example in $x = ax - b$, then it has not been made the subject of the formula.

### Links

This topic could occur in connection with constructing a formula from some physical or commercial situation.

# Inequalities

An inequality involves one of the symbols $<, \leq, >$ or $\geq$. Inequalities can be in one variable or more.

### Methods

The rules for solving an inequality in one variable are the same as those for solving an equation, with one extra rule.

- *When multiplying or dividing by a negative number, reverse the inequality sign.*

If possible, avoid multiplying or dividing by a negative number, as it is very easy to make a mistake.

Inequalities can be shown as a bar on a number line. If the inequality involves $\leq$ or $\geq$ then the endpoint is included, and we use a solid dot. If the inequality involves $<$ or $>$ then the endpoint is not included, and we use a hollow dot. Below are number lines for $x > 2$ and $x \geq 2$.

Inequalities in two variables can be shown on graph paper. If the inequality involves $\leq$ or $\geq$ then the boundary is included and we use a solid line. If it involves $<$ or $>$ then the boundary is not included and we use a broken line. When drawing the boundary line it is often best to find where it crosses the axes. The line $2y + 3x = 6$ crosses the axes at $(0, 3)$ and $(2, 0)$. In the diagrams below, the regions that satisfy the inequalities $2y + 3x \leq 6$ and $2y + 3x < 6$ are shaded.

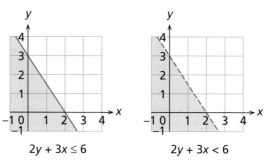

When showing more than one inequality, it is usual to shade *out* the region we *don't* want.

*Examples* Find the smallest integer that satisfies the inequality $x + 3 < 3x - 8$.

$$x - 3x < -3 - 8$$
$$-2x < -11$$

Divide by $-11$. As this is negative, reverse the inequality.

$$x > \tfrac{11}{2} = 5.5$$

The smallest integer satisfying this is 6.

**Note.** We could solve this by collecting the $x$s on the right and the numbers on the left.

$$3 + 8 < 3x - x$$
$$11 < 2x$$
$$5.5 < x \quad \text{(2 is positive, so the inequality sign is unchanged)}$$

This is the same as $x > 5.5$.

---

Solve the inequality $-5 \leq 3x + 1 < 10$. Illustrate your answer on a number line, and list the integers which satisfy the inequality.

Subtract 1, and then divide by 3.

$$-6 \leq 3x < 9$$
$$-2 \leq x < 3 \quad \text{(3 is positive, so the inequality signs are unchanged)}$$

The inequality is shown on the number line. The integers are

$$-2, -1, 0, 1, 2 \quad (-2 \text{ is included, but not } 3)$$

---

Write down the inequalities obeyed in the unshaded region shown.

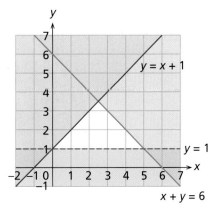

The region is above the line $y = 1$, below the line $x + y = 6$ and below the line $y = x + 1$. The first line is broken, and the other two are solid.

The inequalities are $y > 1$, $x + y \leq 6$ and $y \leq x + 1$.

A textbook costs £8 in paperback and £12 in hardback. A college has £144 to spend on these books. If it buys $x$ paperbacks and $y$ hardbacks, write down an inequality in $x$ and $y$. Illustrate the possible choices on a graph.

The total cost of the books is £$(8x + 12y)$. This is at most £144.
$8x + 12y \leq 144$.

Obviously $x \geq 0$ and $y \geq 0$. The inequality $8x + 12y \leq 144$ simplifies to $2x + 3y \leq 36$. The boundary line goes through (0, 12), which corresponds to buying only hardbacks, and (18, 0), which corresponds to spending all the money on paperbacks.
   In all the inequalities the boundary is included, so they are drawn with solid lines. The result is shown.

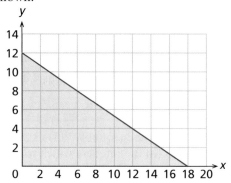

## Exercise on inequalities

You will need:
- graph paper or squared paper
- ruler

1. Solve the following inequalities, illustrating the solutions on a number line.
   a  $3x - 1 \geq 8$
   b  $5 - 2x < 1$
   c  $5 \leq 17 - 3x$
   d  $2x + 5 \leq 3x - 7$
   e  $5x - 2 \geq 2x + 7$
   f  $7 - 3x > 13 - 5x$

2. Solve the following inequalities, in each case listing the integers which satisfy them.
   a  $2 \leq x - 3 < 5$
   b  $-1 < 3x + 4 < 7$
   c  $3 < 5 - 2x \leq 9$
   d  $-8 \leq 5 - 3x < 0$

3. For each of the following, find the least integer which satisfies the inequality.
   a  $3x + 1 \geq 10$
   b  $5x > x + 12$
   c  $2 - 6x < 6 + x$
   d  $9 - 3x \geq 18 - 5x$

4. For each of the following, find the greatest integer which satisfies the inequality.
   a  $3x - 5 < 4$
   b  $2x + 1 \leq 13$
   c  $9 - 2x > 3x + 2$
   d  $3 - 5x \geq 11 - 2x$

5. In the diagram shown, give the inequalities obeyed in the regions shaded
   a  red
   b  blue
   c  green.

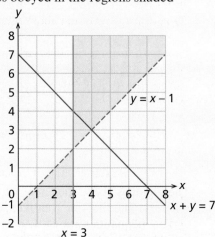

6  Make a copy of this diagram, and on it indicate the regions in which
   a  $x > 1, x + y \leq 5, y \geq x + 1$
   b  $x > 1, x + y \leq 5, y \leq x + 1$
   c  $x < 1, x + y \geq 5, y \geq x + 1$

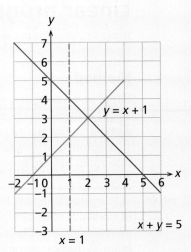

7  Illustrate the following sets of inequalities on graph paper grids.
   a  $y > -1, x + y < 4, y \geq x$
   b  $x > 1, y > 1, 2x + 3y \leq 6, 3x + 2y \leq 6$
   c  $y \geq 0, 3x + 2y \geq 12, 2x + 5y \leq 10$

8  Iris buys $x$ beefburgers at £1.20 each, and $y$ cheeseburgers at £1.50 each. She spends less than £12. Write down an inequality in $x$ and $y$, and illustrate it on a graph.

9  A lorry will carry $x$ cars, of mass 900 kg each, or $y$ vans, of mass 1600 kg each. The total load is at most 36 000 kg. Write down an inequality in $x$ and $y$, and illustrate it on a graph.

10  A firm will employ $x$ skilled workers and $y$ unskilled workers. There must be more skilled than unskilled workers, and at most 20 men can be employed. Write down inequalities in $x$ and $y$ and illustrate them on a graph.

### Question spotting

In many questions the word 'inequality' will occur. In a word problem, watch out for any use of the words 'less than', 'at most', etc.

### Tips

Do not forget the rule about multiplying or dividing by negative numbers. If $x > y$, then $-x < -y$.

Make sure you understand the symbols. If $x \geq 3$ then 3 is included, and it is shown with a solid dot on the number line. If $x < 2$ then 2 is not included and it is shown with a hollow dot. In a two-dimensional inequality, if the boundary is included it is shown with a solid line, and if it is not included it is shown with a broken line.

Watch out for the different ways of expressing inequalities. The following all mean the same, $x \geq 3$.

$x$ is greater than or equal to 3     $x$ is at least 3     $x$ is not less than 3

### Links

An inequality could come as part of a general question on algebra. It could also come as part of a word problem about money, time, etc.

# Linear programming

Linear programming is used to maximise or minimise the value of a certain function.

## Method

Suppose we have two variables, and a set of linear restrictions on the variables. These restrictions can be expressed as inequalities. Illustrate the inequalities on a graph by the method of the previous revision topic.

The function we want to maximise or minimise is the **objective function**. The maximum or minimum value will be at one of the corners of the region.

*Example* A television company has at most 30 hours of time to fill with either documentaries or drama. It costs £240 to produce a minute of a documentary, and £480 for a minute of drama. There is £576000 for the programmes. There must be at least 10 hours of documentaries. Let $x$ and $y$ be the numbers of hours of documentaries and dramas respectively. Find inequalities in $x$ and $y$ and illustrate them.

From previous research it is reckoned that the audiences for a documentary and a drama will be 2 million and 3 million respectively. What allocation of programmes will bring in the greatest audience?

Trivially $x \geq 0$ and $y \geq 0$.
From the number of hours, $x + y \leq 30$.
From the cost restriction,

$$240 \times 60x + 480 \times 60y \leq 576000$$

Divide this by $240 \times 60$.

$$x + 2y \leq 40$$

$x$ must be at least 10.
The restrictions are $x \geq 0$, $y \geq 0$, $x + y \leq 30$, $x \geq 10$, $x + 2y \leq 40$.
The inequalities are shown on the graph.

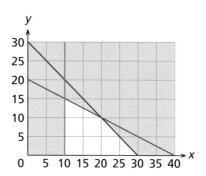

We want to maximise $2x + 3y$. Take a line parallel to $2x + 3y = 0$, for example $2x + 3y = 30$. Slide your ruler up the diagram, keeping it parallel to this line, until it is just about to leave the region.

This happens at (20, 10).

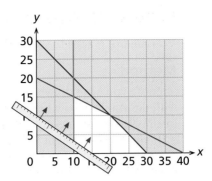

The allocation is 20 hours of documentary, 10 hours of drama.

## Exercise on linear programming

You will need:
- graph paper
- ruler

1. The owner of a furniture shop has $16\,\text{m}^2$ of space to display items. Each table will take $2\,\text{m}^2$, and each chair $1\,\text{m}^2$. A table and a chair cost £120 and £100 respectively, and there is £1200 to spend. Letting $x$ and $y$ represent the numbers of tables and chairs respectively, find inequalities in $x$ and $y$ and illustrate them on a graph.
   If a table and a chair bring in £30 and £20 of profit respectively, find how the owner can maximise the profit.

2. A livestock owner has a choice of two food additives: type A contains 5 units of minerals and 9 units of vitamins per kilogram, and type B contains 10 units of minerals and 6 units of vitamins per kilogram. The animals require at least 200 units of minerals and 180 units of vitamins per day. Letting $x$ kg and $y$ kg be the amounts of type A and type B used per day respectively, find inequalities in $x$ and $y$ and illustrate them on a graph.
   If type A and type B cost £5 and £6 respectively per kilogram, find the least cost of providing the food additives.

3. A school needs to take 120 students on an outing. It can hire either small or large minibuses, which take 10 passengers and 15 passengers respectively. There are only 10 drivers available. The cost of hiring is £40 for a small minibus and £65 for a large one. Find the least possible cost of running the transport for the outing.

### Question spotting

The phrase 'linear programming' may not appear in the question. Be on the lookout for any question which asks you to find the maximum or minimum value of some quantity.

### Tips

Be careful when illustrating the inequalities. If that goes wrong then the whole question goes wrong.
   Read the question carefully, to see whether you are asked to find the optimal value of the objective function, or the values of the variables which give that value.

### Links

Questions on linear programming are long, mark-rich questions. They are likely to appear on their own.

# Graphs

You can use graphs to solve equations and to find formulae.

## Method

Suppose you have drawn the graph of a function. Then the intersection points with the $x$-axis show where the function is zero. You can solve other equations by drawing a suitable straight line and seeing where it crosses the graph.

If you have a straight line graph, its equation can be found using the methods of chapter 7. If you have a graph of the form $y = a^t$, the value of $a$ can be found from a point on the graph.

*Examples* This question is about the graph of $y = x^3 - 2x + 3$.

**a** Complete the table below.

| $x$ | $-2$ | $-1$ | 0 | 0.5 | 1 | 1.5 | 2 |
|---|---|---|---|---|---|---|---|
| $x^3$ | $-8$ | | | 0.125 | | | |
| $-2x$ | 4 | | | | | $-3$ | |
| 3 | 3 | 3 | 3 | | | | |
| $y$ | | | | | | | |

**b** Draw the graph of $y = x^3 - 2x + 3$, for $-2 \leq x \leq 2$.
**c** Use your graph to solve the equation $x^3 - 2x + 3 = 0$.
**d** By drawing a suitable line on your graph, solve the equation $x^3 - x + 1 = 0$.

**a** Fill in the table as shown.

| $x$ | $-2$ | $-1$ | 0 | 0.5 | 1 | 1.5 | 2 |
|---|---|---|---|---|---|---|---|
| $x^3$ | $-8$ | $-1$ | 0 | 0.125 | 1 | 3.375 | 8 |
| $-2x$ | 4 | 2 | 0 | $-1$ | $-2$ | $-3$ | $-4$ |
| 3 | 3 | 3 | 3 | 3 | 3 | 3 | 3 |
| $y$ | $-1$ | 4 | 3 | 2.125 | 2 | 3.375 | 7 |

**b** The graph is shown.

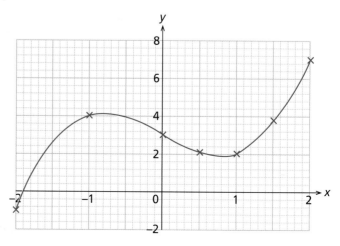

**c** Notice that the graph crosses the $x$-axis at $-1.9$.
   The solution to $x^3 - 2x + 3 = 0$ is $x = -1.9$.
**d** We need to convert $x^3 - x + 1 = 0$ to $x^3 - 2x + 3 = $ some function of $x$.

$$x^3 - x + 1 = 0$$
$$x^3 - x + 3 = 2 \quad \text{(adding 2 to both sides)}$$
$$x^3 - 2x + 3 = 2 - x \quad \text{(subtracting } x \text{ from both sides)}$$

So plot the graph of $y = 2 - x$. This is a straight line, going through $(0, 2)$ and $(2, 0)$. Plot this line on top of the graph.

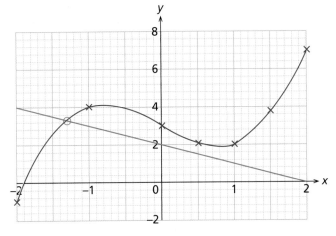

Notice that the graphs cross at $(-1.3, 3.3)$.
The solution of $x^3 - x + 1 = 0$ is $x = -1.3$.

---

In a physics experiment on elasticity, different masses were hung on the end of a spring. In each case the mass $x$ kg and the length $l$ cm of the spring were recorded. The results are below. Plot the points, and find a formula giving $l$ in terms of $x$.

| x (kg) | 0.1 | 0.2 | 0.3 | 0.4 | 0.5 |
|---|---|---|---|---|---|
| l (cm) | 22.4 | 24.8 | 27.2 | 29.6 | 32.0 |

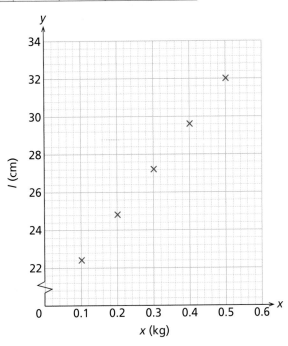

The points are plotted. Notice that they lie exactly on a straight line. We can find the equation of the line as $l = mx + c$. Using the first and the last points, the gradient $m$ of this line is:

$$m = \frac{32.0 - 22.4}{0.5 - 0.1} = \frac{9.6}{0.4} = 24$$

Now use the first point to find $c$.

$$22.4 = 24 \times 0.1 + c$$
$$c = 22.4 - 2.4 = 20$$

The formula is $l = 24x + 20$.

---

An investment was bought for £1000. After $t$ years, the predicted value of the investment is £$V$. The diagram shows a graph of $V$ against $t$. Write down the value of $V$ at $t = 10$. Assuming that the value $V$ is given in terms of $t$ by $V = 1000 \times a^t$, find $a$. Hence predict $V$ at $t = 20$, i.e. after 20 years.

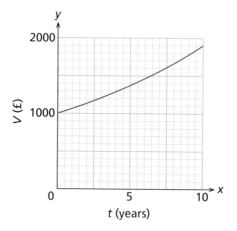

Read off the value from the graph.

At $t = 10$, $V = 1900$. So

$$1900 = 1000 \times a^{10}$$

Hence $a^{10} = 1.9$, giving $a = \sqrt[10]{1.9} = 1.066$.
$a = 1.066$.

Now put $t = 20$ into the equation.

$$V = 1000 \times 1.066^{20} = 3600$$

After 20 years the value is predicted to be £3600.

## Exercise on graphs

**You will need:**
- graph paper
- ruler

**1** Let $y = x^3 + 2x^2 - 11$.

  **a** Complete the table below.

  | x | −2 | −1 | 0 | 0.5 | 1 | 2 |
  |---|---|---|---|---|---|---|
  | $x^3$ | −8 | | | | 1 | |
  | $2x^2$ | 8 | | | | 2 | |
  | −11 | −11 | −11 | | | | |
  | y | | | | | | |

  **b** Draw the graph of $y = x^3 + 2x^2 - 11$, for $-2 \leq x \leq 2$.
  **c** Use your graph to solve the equation $x^3 + 2x^2 - 11 = 0$.
  **d** By drawing a suitable line on your graph, solve the equation $x^3 + 2x^2 + x - 14 = 0$.

**2** Let $y = -x^3 - 2x^2 + x + 3$.

  **a** Complete the table below.

  | x | −3 | −2 | −1 | 0 | 0.5 | 1 | 2 |
  |---|---|---|---|---|---|---|---|
  | $-x^3$ | 27 | | | | | | −8 |
  | $-2x^2$ | | | −2 | | | | |
  | x | | −2 | | | | 1 | |
  | 3 | | | | | | | |
  | y | | | | | | | |

  **b** Draw the graph of $y = -x^3 - 2x^2 + x + 3$, for $-3 \leq x \leq 2$.
  **c** Use your graph to solve the equation $-x^3 - 2x^2 + x + 3 = 0$.
  **d** By drawing a suitable line on your graph, solve the equation $-x^3 - 2x^2 + 5 = 0$.

**3 a** Fill in the table below for the function $y = x^2 - \dfrac{1}{2x}$.

  | x | 0.25 | 0.5 | 0.75 | 1 | 1.5 | 2 |
  |---|---|---|---|---|---|---|
  | $x^2$ | 0.06 | | | | 2.25 | |
  | $-\dfrac{1}{2x}$ | | −1 | | | | −0.25 |
  | y | | | | | | |

  **b** Draw the graph of $y = x^2 - \dfrac{1}{2x}$, for $0.25 \leq x \leq 2$.

  **c** Use your graph to solve the equation $x^2 - \dfrac{1}{2x} = 0$.

  **d** On your graph, draw the graph of the function $y = 2 - x$. Hence solve the equation
  $$2x^3 + 2x^2 - 4x - 1 = 0$$

**4** In an electrical experiment, a fixed current is maintained through a circuit, while the resistance of a component is varied. The voltage, $V$ volts, of the circuit for different values of the resistance, $R$ ohms, is given below. Plot the points, and find a formula giving $V$ in terms of $R$.

| $R$ (ohms) | 5 | 6 | 7 | 8 | 9 |
|---|---|---|---|---|---|
| $V$ (volts) | 4.5 | 5.2 | 5.9 | 6.6 | 7.3 |

**5** A vehicle is decelerating at a steady rate. Its velocity, $v$ m/s, at time, $t$ seconds, is given below. Plot the points, and find a formula giving $v$ in terms of $t$.

| $t$ (s) | 10 | 15 | 20 | 25 | 30 |
|---|---|---|---|---|---|
| $v$ (m/s) | 25 | 22 | 19 | 16 | 13 |

**6** A sapling of height 0.7 m is planted. After $t$ years its height is $h$ m. The diagram shows a graph of $h$ against $t$. Find the height at $t = 2$. Assuming that the height $h$ is given by the formula $h = 0.7 \times a^t$, find $a$. Hence predict the height after three years.

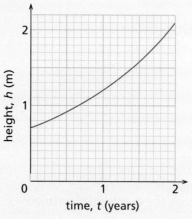

**7** A car is bought for £12 000. After $t$ years its value is £12 000 $\times a^t$. The diagram shows the value $V$ of the car against time $t$. Write down the value after five years, and hence find $a$. Predict the value of the car after seven years.

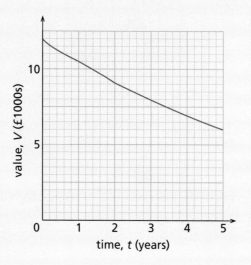

## Question spotting

There should be no difficulty in spotting when a question is asking you to plot a graph and use it to solve equations or to find a formula.

## Tips

When working out the values, be careful with negative values of $x$.

Join the points up by a smooth curve. Do not join the points with straight line segments.

When you are asked to solve an equation in $x$, give only the $x$-coordinate. Do not give the $y$-coordinate.

## Links

Graph questions are long questions, and will probably come on their own.

# Review module 3 – Shape and space

The topics covered in this module are:

- **lines and angles**
- **constructions and loci**
- **Pythagoras' theorem**
- **length, area and volume**
- **dimensions**
- **transformations**
- **similarity**
- **three-dimensional geometry.**

## Lines and angles

This topic deals with the geometry of lines and angles, and of straight line figures.

### Results about angles

- The angles round a point add up to 360°.

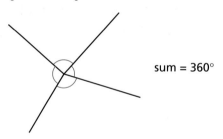

- The angles along a line add up to 180°.

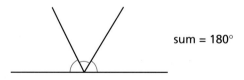

- In the diagram, $AXB$ and $CXD$ are straight lines.

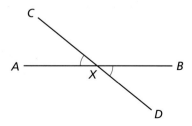

$\angle AXC = \angle BXD$   (vertically opposite)

- In the diagram, the lines $AB$ and $CD$ are parallel.

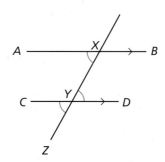

$\angle AXY = \angle DYX$ (alternate)
$\angle AXY = \angle CYZ$ (corresponding)

- The sum of the angles in a triangle is 180°.
- The sum of the angles in a quadrilateral is 360°.
- If a polygon has $n$ sides, the sum of its angles is $180°n - 360°$.
- The sum of the external angles of a polygon is 360°.
- The base angles of an isosceles triangle are equal.

## Results about quadrilaterals

Trapezium:  one pair of sides parallel
no extra properties

Kite: 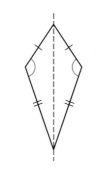 two pairs of adjacent
 sides equal
one pair of angles equal
one line of symmetry

Parallelogram: 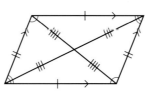 two pairs of opposite
 sides parallel
opposite sides equal
opposite angles equal
diagonals bisect each other
rotational symmetry
 of order 2

Rhombus:

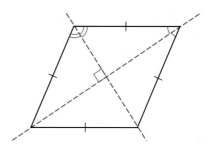

all sides equal
properties as for
parallelograms, and also:
diagonals bisect angles
diagonals perpendicular
two lines of symmetry

Rectangle:

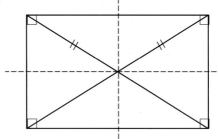

all angles equal to 90°
properties as for
parallelograms, and also:
diagonals equal
two lines of symmetry

Square:

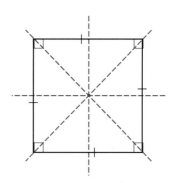

all sides equal, all angles
equal to 90°
properties as for
rectangles and
rhombuses, and also:
four lines of symmetry
rotational symmetry of
order 4

*Examples*  In the diagram, $AB$ and $CD$ are parallel. $\angle ABC = 32°$ and $\angle BDC = 78°$. Find
 **a** $\angle DCB$   **b** $\angle CBD$

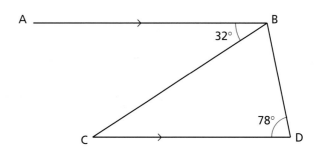

**a** $\angle DCB = \angle ABC$   (alternate)
   $\angle DCB = 32°$.
**b** $\angle CBD = 180° - \angle DCB - \angle BDC$
   $\angle CBD = 70°$.

In the diagram, $AB = AD$ and $CB = CD$. $\angle DAB = 138°$ and $\angle DEA = 25°$. Find $\angle ABC$.

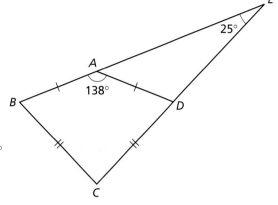

$\angle DAE = 180° - 138° = 42°$
  (angles on a straight line)
$\angle ADE = 180° - 42° - 25° = 113°$
  (angles in a triangle)
$\angle ADC = 180° - 113° = 67°$
  (angles on a straight line)
$ABCD$ is a kite, so $\angle ABC = \angle ADC$.
$\angle ABC = 67°$.

---

$ABCDEF$ is a regular hexagon. Find the angles
**a** $\angle ABC$ **b** $\angle DBC$

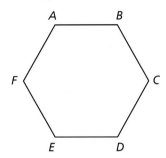

**a** The sum of the angles is $180° \times 6 - 360° = 720°$. The hexagon is regular, so divide this by 6.
$\angle ABC = 120°$.
**b** Similarly, $\angle BCD = 120°$. $BCD$ is isosceles, so
$\angle DBC = \angle BDC = \frac{1}{2}(180° - 120°)$.
$\angle DBC = 30°$.

# Exercise on lines and angles

**1** In the diagram, $AB$ and $CD$ are parallel. $\angle BAE = 28°$ and $\angle DCE = 37°$. Find $\angle AEC$.

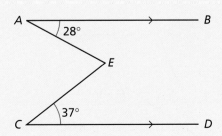

**2** In the diagram, $AB = BC = BD$ and $AC$ and $DE$ are parallel. $\angle BAC = 56°$. Find
  **a** $\angle CBD$        **b** $\angle DCB$        **c** $\angle EDC$

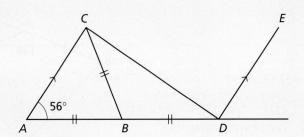

**3** In the diagram, $AB = CB = CD$. $\angle EDC = 132°$. Find $\angle BAC$.

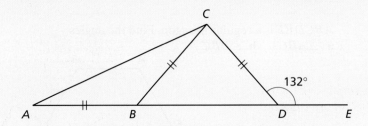

**4** In the diagram, $AB$ and $CD$ are parallel. $\angle CDE = 22°$ and $\angle CED = 104°$. $CD$ is the bisector of $\angle BCE$. Find $\angle ABC$.

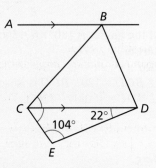

**5** One angle of a parallelogram is $73°$. Find the other angles.
**6** $ABCD$ is a rhombus, and $\angle ACD = 52°$. Find $\angle ABC$.
**7** $ABCD$ is a parallelogram. The bisectors of the angles at $C$ and $D$ meet at $X$. Find $\angle DXC$.

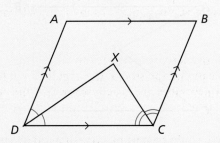

**8** In the diagram, *ABCD* is a straight line with *AB* = *CD* = *BE*. *BEFC* is a parallelogram. ∠*EAB* = 34°. Find ∠*AGD*.

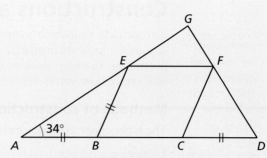

**9** In the diagram, *ABCD* is a rhombus and *ABE* is an equilateral triangle. ∠*ADC* = 72°. Find ∠*ADE*.

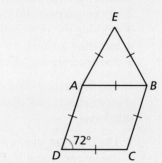

**10** In the diagram, △*ABC* is equilateral and *ABDE* and *BCFG* are squares. Find ∠*AEF*.

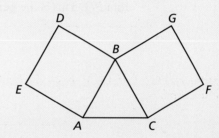

**11** Find the internal angle of a regular 12-sided figure.
**12** The internal angle of a regular polygon is 8 times as large as the external angle. Find the number of sides of the polygon.
**13** The internal angle of a regular polygon is 150° greater than the external angle. Find the number of sides of the polygon.
**14** *ABCDE* is a regular pentagon. *X* is the midpoint of *CD*. Find ∠*CAX*.
**15** *ABCDEFG* is a regular heptagon. *GA* and *CB* produced meet at *H*. Find ∠*AHB*.

### Question spotting

In this sort of question you are often asked to find angles without being given any lengths (unlike questions involving trigonometry).

### Tips

Always give a reason for a result.
   You can find the sum of the angles of a polygon. But you cannot find the angles themselves, unless you know that the polygon is regular.

### Links

The topic could be part of a question involving other parts of geometry, such as similarity or Pythagoras' theorem.

## Constructions and loci

Geometrical instruments consist of a ruler, compasses and a protractor. In this section we look at constructions using ruler and compasses only, i.e. not using a protractor. In many cases we construct the **locus** of a point, i.e. the path which a point traces out when it moves according to a certain rule.

> Note the distinction between a line and a line segment. The line *l* may stretch infinitely in both directions. The line segment AB only goes from A to B.

### Methods of construction

The basic constructions which require ruler and compasses only are as follows:

1. drawing a perpendicular from a point to a line,
2. drawing a perpendicular from a point on a line,
3. drawing the perpendicular bisector of a line,
4. bisecting an angle,
5. constructing angles of 60°, 30°, 45°, etc.

In the instructions below, an arc drawn from a point *P* is drawn with the point of the compasses at *P*. Keep the separation of the compasses constant, unless told otherwise.

1. The diagram shows a line and a point *P* above it. Draw an arc from *P*, cutting the line at *A* and *B*. Draw arcs from *A* and *B*, meeting at *Q* below the line. Join *PQ*. This is the perpendicular.

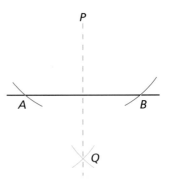

2. The diagram shows a line and a point *P* on it. Draw an arc from *P*, cutting the line at *A* and *B*. Widen the compasses and draw arcs from *A* and *B*, meeting at *Q* above the line. Join *PQ*. This is the perpendicular.

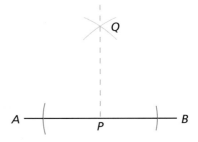

**Note.** If *P* is at the end of the line, just extend the line.

**3** The diagram shows a line segment $AB$. Draw arcs from $A$ and $B$, meeting at $C$ above the line and $D$ below the line. Join $CD$. This is the perpendicular bisector.

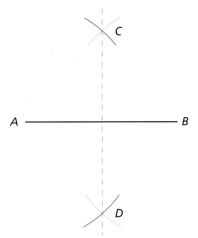

**4** The diagram shows lines $l$ and $m$ meeting at $O$. Draw an arc from $O$, meeting the lines at $A$ and $B$. Draw arcs from $A$ and $B$, meeting at $C$. Join $OC$. This is the bisector of the angle.

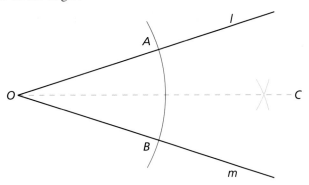

**5** The diagram shows a line segment $AB$. Draw an arc from $A$, cutting the segment at $C$. Draw an arc from $C$, cutting the first arc at $D$. Join $AD$. $\angle DAC = 60°$.

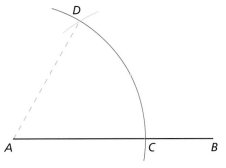

Bisect this angle (see construction 4) to obtain an angle of $30°$.
Bisect a right angle (see constructions 1, 2 above) to obtain an angle of $45°$.

## Loci

1 If a point moves a fixed distance from a fixed line, then its locus consists of the lines parallel to the fixed line.

2 If a point moves a fixed distance from a fixed point, then its locus is the circle with centre at the fixed point.

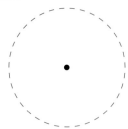

3 If a point moves so that it is equidistant from points $A$ and $B$, then its locus is the perpendicular bisector of $AB$. See construction 3 above.
4 If a point moves so that it is equidistant from lines $l$ and $m$, then its locus is the bisector of the angle between $l$ and $m$. See construction 4 above.

The **circumcircle** of a triangle goes through its vertices. Its centre (the circumcentre) is the intersection of the perpendicular bisectors of the sides.

The **incircle** of a triangle touches its sides. Its centre (the incentre) is the intersection of the bisectors of the angles.

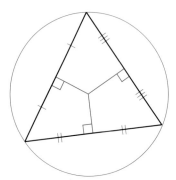 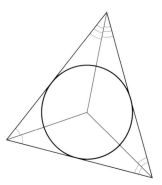

*Example*  The diagram shows a rectangle $ABCD$ with $AB = 4$ cm and $AD = 3$ cm.

Make a copy of the diagram and construct the bisector of the angle $ABC$. Construct the perpendicular bisector of $AD$. These lines meet at $X$. Measure the distance $BX$.

Perform these constructions as outlined above. The construction lines are shown. Finally measure $BX$.

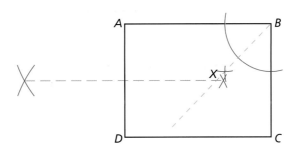

$BX = 2.1$ cm.

# Exercise on constructions and loci

**You will need:**
- compasses
- ruler

1 Make a copy of each diagram and construct:

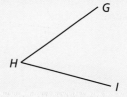

   **a** the perpendicular from $P$ to $AB$        **b** the perpendicular to $CD$ from $Q$

   **c** the perpendicular bisector of $EF$        **d** the bisector of $\angle GHI$.

2 Construct a square $ABCD$ of side 5 cm. Bisect the line $DC$ and the angle $\angle ABC$. Show that these bisectors meet on the diagonal $AC$.

3 Construct an equilateral triangle $ABC$ of side 5 cm. Bisect the angles at $B$ and at $C$. Let these bisectors meet at $X$. Measure $AX$.

4 Construct angles of
   **a** $22\frac{1}{2}°$              **b** $15°$              **c** $75°$

5 Construct a triangle $ABC$ with $\angle ABC = 90°$, $AB = 4$ cm and $BC = 5$ cm. Find the points which are equidistant from $A$ and $C$ and are 2 cm from $B$.

**6** The diagram shows a map of a harbour. To enter the harbour, a ship should be steered so that it is equidistant from $X$ and $Y$. Make a copy of the map and construct the ship's course.

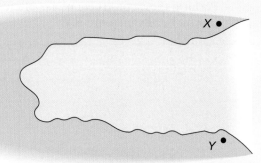

**7** The diagram shows a map of three cities, to a scale of 1 cm per 100 km. A transmitting station is to be sited so that it is equidistant from the three cities. Make a copy of the diagram. By construction, find the position of the station, and find its distance from the cities.

A •

• B

C •

**8** Construct a triangle with sides 4 cm, 5 cm and 6 cm. Construct the circumcircle and the incircle. Measure their radii.

**9** Construct a triangle $ABC$ with $AB = 6$ cm, $\angle CAB = 60°$ and $\angle CBA = 45°$. Construct the circumcircle and the incircle, and measure their radii.

**10** The diagram shows a map, to a scale of 1 cm to 10 m. A treasure is buried on a spot which is 15 m from the wall, and equidistant from the trees at $T$ and $S$. Make a copy of the map and mark on it the position of the treasure.

S •

• T

**11** Construct a rectangle $ABCD$ with $AB = 4$ cm and $AD = 5$ cm. Make a total of three copies of this rectangle, to answer the following three questions.
  **a** Indicate, by shading, those points nearer to $AB$ than to $AD$.
  **b** Indicate, by shading, those points nearer to $A$ than to $B$ and within 3 cm of $C$.
  **c** Indicate, by shading, those points nearer to $AB$ than to $BC$ and within 1 cm of $AD$.

**12** Construct a square $ABCD$ of side 4 cm.
  **a** Construct the locus of points inside the square which are a distance of 1 cm from the nearest side of the square.
  **b** Construct the locus of points outside the square which are a distance of 1 cm from the nearest side of the square.

### Question spotting

The words 'construct' or 'locus' may or may not appear in the question. You may be given a description of a shape or path, and asked to draw it using ruler and compasses only.

### Tips

For a construction question, you are expected to do the construction using ruler and compasses only. You would not be allowed, for example, to bisect an angle by measuring it with a protractor, halving the number and then drawing the half angle. But you can use your protractor to *check* the result.

Leave your construction lines in the diagram, even if they look messy. They show the examiner what you have done. In particular, they show that you have done it by construction, rather than by measurement and drawing.

### Links

This topic is likely to occur on its own, or with other geometric topics.

# Pythagoras' theorem

If you know two sides of a right-angled triangle, then you can use Pythagoras' theorem to find the third side.

### Definition

- Suppose $ABC$ is a right-angled triangle, with $\angle BAC = 90°$.
  The **hypotenuse** of the triangle is always the longest side, in this case $BC$.

### Result

The square on the hypotenuse is equal to the sum of the squares on the other two sides.

$$BC^2 = AB^2 + AC^2$$

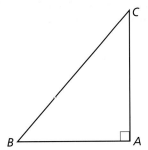

*Examples*  In $\triangle PQR$, $\angle QRP = 90°$, $QR = 8.7$ cm and $RP = 11.2$ cm. Find $QP$, correct to one decimal place.

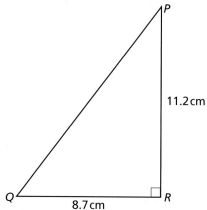

Draw a sketch of the triangle. Note that $QP$, the side we want to find, is the hypotenuse.

So use the theorem directly.

$$QP^2 = 8.7^2 + 11.2^2 = 201.13$$

Hence $QP = \sqrt{201.13} = 14.2$ cm.

---

A straight rope connects the top of a pole with the ground. The pole is 6 m high, and the length of the rope is 10 m. How far is the end of the rope from the foot of the pole?

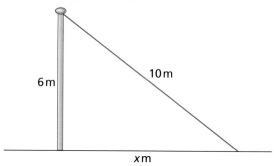

From the sketch, the length of the rope is the hypotenuse. The horizontal distance is the unknown. Let this distance be $x$ m. By the theorem

$$10^2 = 6^2 + x^2$$
$$x^2 = 100 - 36 = 64$$
$$x = \sqrt{64} = 8$$

The end of the rope is 8 m from the foot of the pole.

**Note.** the answer came out to exactly 8 m, as 64 is a perfect square. You could be expected to do this question without a calculator.

The sides of a right-angled triangle are $(2x + 1)$ m, $3x$ m and $(5x - 5)$ m, where the hypotenuse is the last of the three. Find $x$.

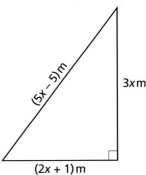

Notice that you are not 'led through' this question. You have to set up the equation for yourself and then solve it. Using Pythagoras' theorem:

$$(2x + 1)^2 + (3x)^2 = (5x - 5)^2$$
$$4x^2 + 4x + 1 + 9x^2 = 25x^2 - 50x + 25$$
$$12x^2 - 54x + 24 = 0$$
$$2x^2 - 9x + 4 = 0$$
$$(2x - 1)(x - 4) = 0$$

Hence $x = \frac{1}{2}$ or $x = 4$.

The solution $x = \frac{1}{2}$ gives a negative value for the hypotenuse, $5x - 5$. Therefore $x = 4$.

---

A triangle has sides 9 cm, 9 cm and 8 cm. Find the height of the triangle, leaving your answer in terms of a square root.

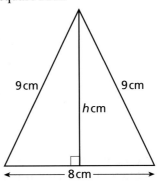

This triangle is isosceles but not right-angled. But it can be converted to two right-angled triangles by dropping a perpendicular, which is the height of the triangle. If this height is $h$ cm

$$h^2 + 4^2 = 9^2$$
$$h^2 = 81 - 16 = 65$$

The height of the triangle is $\sqrt{65}$ cm.

# REVIEW MODULE 3 – SHAPE AND SPACE

## Exercise on Pythagoras' theorem

1 Find the unknown sides in the diagrams below. Where relevant leave your answers in terms of a square root.

**a** 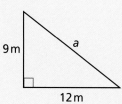    **b** 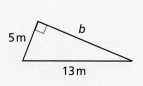    **c** 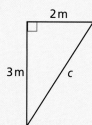    **d**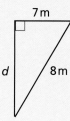

2 A rectangle has sides 30 m and 40 m. Find the length of the diagonal of the rectangle.
3 I walk 10 miles, so that I am 8 miles north of my starting point. How far east or west am I from my starting point?
4 In the quadrilateral shown, $\angle BAC = \angle BDC = 90°$, $AB = 2$ cm, $AC = \sqrt{3}$ cm and $BD = \sqrt{5}$ cm. Find $CD$. Leave your answer in terms of a square root.

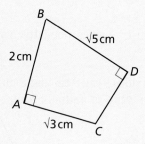

5 In $\triangle ABC$, $\angle BAC = 90°$, $AB = x$ cm, $AC = (2x + 2)$ cm and $BC = (3x - 2)$ cm.
  **a** Form an equation in $x$.
  **b** Solve the equation to find $x$.
6 The three sides of a right-angled triangle are $(x - 3)$ cm, $(x - 2)$ cm and $(11 - x)$ cm, where the last named is the hypotenuse. Find $x$.
7 The lines on a graph paper are 1 cm apart. What are the distances between the following pairs of points? Where relevant leave your answer in square root form.
  **a** (1, 1) and (4, 5)      **b** (3, 7) and (1, −3)      **c** (−4, 7) and (4, −8)

Where relevant, give your answers correct to three significant figures.
8 Find the unknown lengths in the triangles below.

**a** 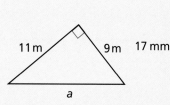    **b** 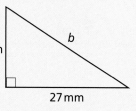    **c** 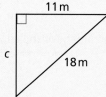    **d**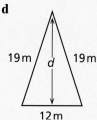

9 In the diagram, $\angle ABC = \angle ADC = 90°$, $AB = 3\,\text{m}$, $BC = 5\,\text{m}$ and $CD = 4\,\text{m}$.
  a Find $AC$.
  b Find $AD$.

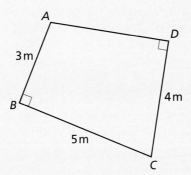

10 In the diagram, $\angle PQR = \angle SPR = 90°$, $PQ = 4.2\,\text{m}$, $QR = 3.7\,\text{m}$ and $SP = 2.9\,\text{m}$. Find $SR$.

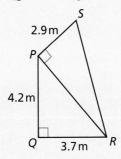

11 The diagonal of a rectangular field is 200 m, and the length and the width are in the ratio 7:5. Let the width be $x$ m.
  a Find the lengths in terms of $x$.
  b Form an equation in $x$.
  c Solve the equation to find the width.
12 In $\triangle PQR$, $\angle PRQ = 90°$, $PQ = (3x + 2)\,\text{cm}$, $PR = (4x - 5)\,\text{cm}$ and $RQ = (x + 3)\,\text{cm}$. Find $x$.
13 The hypotenuse of a right-angled triangle is 100 cm, and the difference between the shorter sides is 68 cm. Find the shortest side of the triangle.
14 The sides of a triangle are 23 m, 23 m and 18 m. Find the height of the triangle.
15 The two sides of a stepladder are 2.8 m long. The feet are 1.4 m apart. How high is the top of the stepladder above the ground?

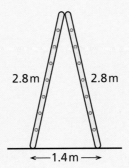

### Question spotting

The word 'Pythagoras' is unlikely to occur in the question. You are expected to recognise the situation in which it can be applied. So watch out for mention of the phrase 'right-angled', or for any situation in which it is clear there is a right-angled triangle or an isosceles triangle. If you are told two of the sides, then you can use the theorem to find the third side.

### Tips

Make sure you distinguish between the two cases of the theorem:

- given the shorter sides, find the hypotenuse
- given the hypotenuse and one shorter side, find the other shorter side.

Don't round numbers too soon. In the first example above, you were asked to give the answer correct to one decimal place. Leave the rounding till the end – don't round the intermediary steps, otherwise the final answer may be inaccurate.

### Links

Questions on Pythagoras' theorem are often linked with the following.

1 Trigonometry: in questions about triangles, you could also be asked to find the angles.
2 Algebra: some of the examples above lead to quadratic equations.
3 Square roots: the sides of a triangle might be given in terms of square roots.
4 Circles: there are many results about circles which involve right angles. For example, the tangent to a circle is perpendicular to the radius at the point of contact.

## Length, area and volume

### Results

| shape | perimeter | formula | |
|---|---|---|---|
| rectangle | twice sum of breadth and height | $2(b + h)$ | |
| circle | $2\pi \times$ radius | $2\pi r$ | |
| arc | perimeter of circle $\times \dfrac{\theta}{360}$ | $\pi r \dfrac{\theta}{180}$ | |

# Length, area and volume

| shape | area | formula | |
|---|---|---|---|
| rectangle | base × height | $bh$ | |
| parallelogram | base × height | $bh$ | |
| triangle | ½ base × height<br>½ product of sides × sine of enclosed angle | $\frac{1}{2}bh$<br>$\frac{1}{2}ab\sin C$ | |
| trapezium | average of parallel sides × height | $\frac{1}{2}(a+b)h$ | |
| circle | π × radius squared | $\pi r^2$ | |
| sphere (surface) | 4π × radius squared | $4\pi r^2$ | |
| sector | area of circle × $\frac{\theta}{360}$ | $\pi r^2 \frac{\theta}{360}$ | |
| segment | area of sector − area of triangle | $\pi r^2 \frac{\theta}{360} - \frac{1}{2}r^2 \sin\theta$ | |

# REVIEW MODULE 3 – SHAPE AND SPACE

| shape | volume | formula | |
|---|---|---|---|
| cuboid | breadth × height × depth | $bhd$ | |
| prism | base area × height | $Ah$ | |
| pyramid | ⅓ base area × height | $\frac{1}{3}Ah$ | |
| cylinder | base area × height | $\pi r^2 h$ | |
| cone | ⅓ base area × height | $\frac{1}{3}\pi r^2 h$ | |
| sphere | $\frac{4}{3}\pi$ × radius cubed | $\frac{4}{3}\pi r^3$ | |

The **density** of a substance is its mass divided by its volume.

*Examples*   A cylinder has height 5 m and radius 7 m. What is its volume?

Use the formula, putting $h = 5$ and $r = 7$.

$$V = \pi \times 7^2 \times 5 = 245\pi$$

The volume is 770 m³.

The area of a circle is 93 cm². Find its radius.

Use the formula for the area of a circle. Then solve to find $r$.

$$93 = \pi r^2$$
$$r^2 = \frac{93}{\pi}$$
$$r = \sqrt{\frac{93}{\pi}}$$

The radius is 5.44 cm.

---

The diagram shows a shed.
Find
a the area of a side wall
b the volume of the shed.

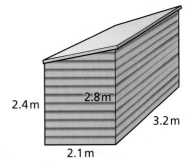

a The side of the shed is a trapezium. The parallel sides are 2.4 m and 2.8 m, and their distance apart is 2.1 m. Use the formula

$$A = \tfrac{1}{2} \times 2.1(2.4 + 2.8)$$
$$= 1.05 \times 5.2$$

The area is 5.46 m².

b The shed is a prism, with length 3.2 m and cross-sectional area 5.46 m².

$$V = 3.2 \times 5.46$$

The volume of the shed is 17.5 m³.

---

A regular hexagon is inscribed in a circle of radius 12 cm. Find the area between the hexagon and the circle.

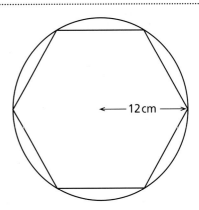

The area consists of six segments. The angle subtended by the segment is 360° ÷ 6, i.e. 60°. Use the formula for the area of a segment.

$$A = \pi r^2 \frac{\theta}{360} - \tfrac{1}{2} r^2 \sin\theta$$

$$= \pi 12^2 \times \tfrac{60}{360} - \tfrac{1}{2} \times 12^2 \sin 60°$$
$$= \pi 24 - 72 \times 0.8660$$
$$= 13.04$$

Finally multiply by 6.
The area is 78.3 cm².

A solid sphere of radius 3 cm has mass 1033 gram. Find its density in kg/m³.

The mass is 1.033 kg. The volume in m³ is

$$\tfrac{4}{3}\pi 0.03^3 = 0.000113$$

Divide the mass by the volume.
The density is 9130 kg/m³.

## Exercise on length, area and volume

1 Find the area of a triangle:
  a with base 7 cm and height 3 cm
  b with $AB = 16$ m, $BC = 23$ m and $\angle ABC = 42°$.
2 Find the volume of each of the following.
  a A cylinder with height 5.3 cm and radius 0.8 cm
  b A sphere with radius 1.5 m
  c A cone with height 38 cm and radius 17 cm
3 A cuboid with a square base of side 16 m has a volume of 1024 m³. Find its height.
4 A cylindrical can has height 10 cm and volume 980 cm³. Find its radius.
5 A sphere has surface area 1098 mm². Find its radius and its volume.
6 A sphere has volume 34 cm³. Find its surface area.
7 An unsharpened pencil of length 12 cm has a cross-section which is a hexagon of side 0.5 cm. Find
  a the area of the hexagon
  b the volume of the pencil.

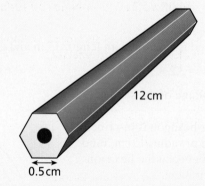

8 The diagram shows a hut. Find
  a the area of the end wall
  b the volume of the hut.

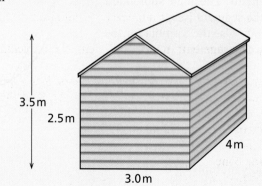

9 The diagram shows a hexagon cut out of a square piece of cardboard. Find the percentage of the cardboard that is not used.

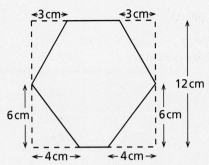

10 A window is in the shape of a semicircle on top of a rectangle. If the window is made of glass 1.5 mm thick, find the volume of the glass.

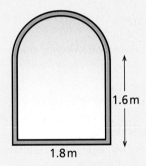

11 The Earth is a sphere of radius $6.4 \times 10^6$ m. Find its volume and surface area, leaving your answers in standard form.
12 A certain planet has volume $2.4 \times 10^{21}$ m³. Find its surface area, leaving your answer in standard form.
13 The diagram shows a 14 cm length of wire bent into a sector.

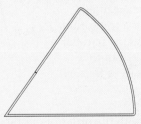

   a If the radius of the circle is 5 cm, find the angle of the sector.
   b If the angle of the sector is 70°, find the radius of the circle.
14 The internal radius of a copper pipe is 2.8 cm and the external radius is 3 cm. Find the volume of copper in 8 m of the pipe.
15 A hollow rubber ball is a sphere with external radius 6.4 cm. If the thickness of the rubber is 4 mm, find the volume of rubber.
16 A cylinder containing water has radius 5 cm. A sphere of radius 3 cm is put into the water, the sphere is immersed and the water does not overflow out of the cylinder. Find the rise in the level of the water.
17 A cone has base radius 6 cm and height 11 cm.
   a Find the volume of the cone.
   b Find the slant height of the cone.
   c Find the curved surface area of the cone.

**18** The top of a cone is cut off. The remaining frustum has height 2 m, top radius 0.8 m and bottom radius 1.6 m.
  **a** Find the height of the original cone.
  **b** Find the volume of the frustum.
  **c** Find the total surface area of the frustum.

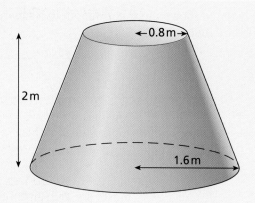

**19** A solid cylinder has height 4 m and radius 3 m. Its mass is 600 000 kg. Find the density of the material from which it is made.

**20** A spherical rubber ball has internal radius 4.3 cm and external radius 4.5 cm. The rubber has density 0.95 grams per cm³. Find the mass of the ball.

**21** A solid sphere made from material of density 835 kg/m³ has mass 83 kg. Find the radius of the sphere.

### Question spotting

Often a question will ask you to find an area or volume. It may involve density, and ask you to find a mass.

### Tips

Of all the shapes mentioned, the one you are least likely to spot is the trapezium. The picture for a trapezium normally shows the parallel sides horizontal, but they could just as well be vertical.

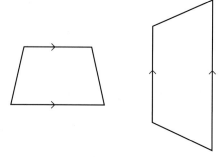

Be careful with the order of operations. For example:

$A = \pi r^2$  to find $A$, given $r$: square $r$ first, *then* multiply by $\pi$
$A = \pi r^2$  to find $r$, given $A$: divide $A$ by $\pi$ first, *then* take the square root

Be careful with accuracy, especially when one part of a question depends on the answer to a previous part of the question. Keep *all* the answer on your calculator.

Be sure that your answer includes units. For a volume, give the answer as, for example, 5.2 cm³, not as 5.2 or as 5.2 cm.

Give your answers to an appropriate degree of accuracy.

### Links

Many other topics could be linked with a question on lengths, areas and volumes. Questions could involve Pythagoras' theorem, trigonometry, accuracy, powers, standard form, estimation and so on.

## Dimensions

By looking at an expression, you can tell whether it represents length, area or volume.

### Results

- The sum or difference of two lengths represents length.
- The product of two lengths represents area.
- The product of three lengths represents volume.
- When a length, area or volume is multiplied by a dimensionless constant, such as 4, $\pi$ or $\frac{2}{3}$, it remains a length, area or volume.
- When a formula involves the addition of two expressions, those expressions must have the same dimensions.

Look at the formula for the area of a trapezium, in the previous section.

$$A = \tfrac{1}{2}(a+b)h$$

The $\frac{1}{2}$ is a dimensionless constant. Ignore it.
    The sum of $a$ and $b$ is the sum of two lengths, hence is a length.
    So the formula is the product of two lengths, hence represents area.

*Example*    Which of length, area or volume could the following represent? The letters $a$ and $b$ represent length.
**a** $\pi ab$    **b** $\tfrac{1}{4}a^2(a+b)$

**a** $\pi$ is a dimensionless constant, so it can be ignored. We now have two lengths multiplied together.
So $\pi ab$ could represent area.
**b** Ignore the factor of $\tfrac{1}{4}$. The sum of $a$ and $b$ is a length. This is multiplied by $a^2$, so in all we have three lengths multiplied together.
So $\tfrac{1}{4}a^2(a+b)$ could represent volume.

## Exercise on dimensions

**1** Which of length, area or volume could the following represent? The letters $x$, $y$ and $z$ represent lengths. The numbers are dimensionless constants.
  **a** $3x + \pi y$      **b** $x(y+z)$      **c** $3x^2 - \pi yz$      **d** $\pi(x+y+z)$
  **e** $\tfrac{4}{3}\pi x^2 y$      **f** $\pi z^3 + 3xyz$      **g** $\sqrt{x^2 + y^2}$      **h** $\dfrac{4\pi x^2 y}{z}$

**2** Which of the expressions below could represent area? The letters $a$, $b$ and $c$ represent lengths. The numbers are dimensionless constants.
  **a** $3a(b-c)$      **b** $\pi(2a + 3b)$      **c** $ab \times ac$
  **d** $a\sqrt{b^2 + c^2}$      **e** $\dfrac{a^3}{2b}$

**3** Which of the formulae below must be incorrect? The letters $r$, $h$ and $l$ represent lengths. The numbers are dimensionless constants.
  **a** $A = \pi r^2 + hl$
  **b** $V = 2\pi r(rh - 3l)$
  **c** $L = \sqrt{rh} - hl$
  **d** $V = (\tfrac{1}{5}h) \times (\tfrac{1}{7}rl)$

### Question spotting
Many questions involve formulae, but for this topic you are not asked to evaluate a formula or simplify it, merely to say what it represents.

### Tips
Ignore constants like $\pi$, 3, $\frac{4}{5}$ and so on.
  When you add or subtract terms of a certain dimension, the result has the same dimensions. When you multiply or divide terms, the dimensions do change.

### Links
A question about dimensions is most likely to occur by itself.

# Transformations
## Definitions and methods
The types of transformations are:

| | |
|---|---|
| translation | moves all points in a certain direction |
| reflection | reflects all points in a fixed mirror line |
| rotation | turns all points about a fixed point through a fixed angle |
| enlargement | enlarges points by a fixed scale factor from a fixed point |

- A **combination** of transformations is when one transformation is done after another.
- The **inverse** of a transformation is the transformation which returns all points to their original position.
- Suppose $AB$ is enlarged to $A'B'$. The centre of enlargement $X$ is the intersection of $AA'$ and $BB'$.

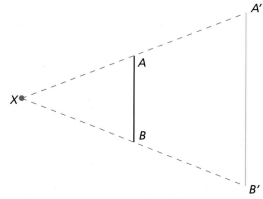

- Suppose $AB$ is rotated to $A'B'$. The centre of rotation $C$ is the intersection of the perpendicular bisectors of $AA'$ and $BB'$.

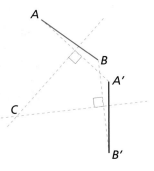

- A shape is symmetrical about a line if it is unchanged after reflection in the line.
- A shape is symmetrical about a point if it is unchanged after rotation about the point. The order of symmetry is the number of part turns needed to get the shape back to its original position.

*Examples*  The diagram shows a triangle with vertices at $A(-2, -1)$, $B(-3, 2)$ and $C(-1, 1)$.

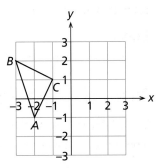

**a** $ABC$ is rotated through 90° clockwise about the origin to $A'B'C'$. Draw $A'B'C'$ on the diagram and write down its coordinates.

**b** $A'B'C'$ is reflected in the $x$-axis to $A''B''C''$. Draw $A''B''C''$ on the diagram and write down its coordinates.

**c** What single transformation will take $A''B''C''$ to $ABC$?

**a** For each vertex of $ABC$, join to the origin and rotate through 90°. The image is shown in red.

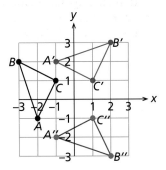

The new vertices are $A'(-1, 2)$, $B'(2, 3)$, $C'(1, 1)$.

**Note.** The coordinates are interchanged and the new $y$-coordinate is multiplied by $-1$.

**b** For each vertex of $A'B'C'$, reflect in the $x$-axis. The image is shown in blue. The new vertices are $A''(-1, -2)$, $B''(2, -3)$, $C''(1, -1)$.

**Note.** The $x$-coordinates remain the same, and the $y$-coordinates are multiplied by $-1$.

**c** The triangles $ABC$ and $A''B''C''$ are symmetrically placed on either side of the line $y = x$.
Reflection in $y = x$ transforms $A''B''C''$ to $ABC$.

In the diagram, $\triangle ABC$ has been enlarged to $\triangle PQR$. Find the centre of enlargement and the scale factor.

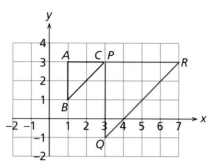

Join $AP$ and $BQ$. Notice that they cross at $(-1, 3)$. $AB$ is 2 units and $PQ$ is 4 units.

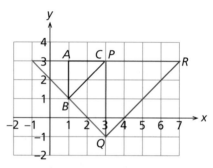

The centre is $(-1, 3)$ and the scale factor is 2.

## Exercise on transformations

You will need:
- graph paper or squared paper
- ruler

1 The triangle shown has vertices at $A(1, 1)$, $B(3, 2)$, $C(3, 1)$.
  a $ABC$ is reflected in the $x$-axis to $A'B'C'$. Draw the new triangle and write down the coordinates of its vertices.
  b $A'B'C'$ is reflected in the line $y = x$ to $A''B''C''$. Draw the new triangle and write down the coordinates of its vertices.
  c Find the single transformation which will take $ABC$ to $A''B''C''$.

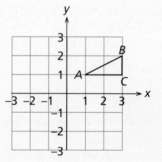

2 The diagram shows a triangle $T$.
  a Triangle $T$ is transformed to triangle $T'$ by a rotation about the origin of $90°$ anti-clockwise. Draw $T'$ and write down the coordinates of its vertices.
  b Triangle $T'$ is transformed to triangle $T''$ by a reflection in the $x$-axis. Draw $T''$ and write down the coordinates of its vertices.
  c Find the single transformation which will take $T$ to $T''$.

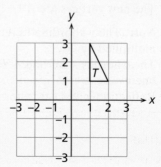

3 Plot a triangle at $A(1, 0)$, $B(3, 2)$ and $C(1, 2)$.
  a $ABC$ is enlarged from $(-1, 2)$ by a scale factor of $-\frac{1}{2}$. Plot the enlarged triangle $A'B'C'$ and write down the coordinates of its vertices.
  b $A'B'C'$ is enlarged from $(-2, 1)$ by a scale factor of $-2$. Plot the enlarged triangle $A''B''C''$ and write down the coordinates of its vertices.
  c What single transformation will take $ABC$ to $A''B''C''$?
4 The diagram shows a shape $S$.

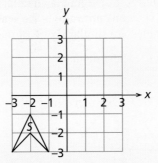

  a Draw the image $S'$ of $S$ under a reflection in the line $y = -x$.
  b Draw the image $S''$ of $S'$ under a reflection in the $y$-axis.
  c Find the single transformation which will take $S''$ to $S$.
5 A triangle $T$ has vertices at $(1, -1)$, $(2, 2)$ and $(2, -2)$.
  a Plot $T$ on graph paper.
  b $T$ is reflected to $T'$ in the line $x = 3$. Draw $T'$ and write down the coordinates of its vertices.
  c $T'$ is reflected to $T''$ in the line $x = 1$. Draw $T''$ and write down the coordinates of its vertices.
  d Describe the single transformation which will take $T$ to $T''$.
6 A triangle has vertices at $A(3, 1)$, $B(2, 4)$ and $C(4, 1)$.
  a Plot $ABC$ on graph paper.
  b $ABC$ is rotated about the origin by $180°$ to $A'B'C'$. Draw $A'B'C'$ on the same graph paper, and write down the coordinates of its vertices.
  c $ABC$ is reflected in the $y$-axis to $A''B''C''$. Draw $A''B''C''$ on the same graph paper, and write down the coordinates of its vertices.
  d Describe the transformation which will take $A''B''C''$ to $A'B'C'$.
7 A triangle with vertices at $(2, 0)$, $(2, 2)$ and $(1, 2)$ is enlarged to a triangle with vertices at $(0, -6)$, $(0, 0)$ and $(-3, 0)$. Plot the triangles and find the centre and scale factor of the enlargement.
8 A triangle with vertices at $(1, 1)$, $(1, 2)$ and $(3, 2)$ is rotated to a triangle with vertices at $(-1, 1)$, $(0, 1)$ and $(0, -1)$. Plot the triangles and find the centre and angle of the rotation.

## Question spotting

In any question involving transformations one of the words 'translation', 'reflection', 'rotation' or 'enlargement' should appear.

## Tips

The order of transformations is important. X followed by Y may be different from Y followed by X.
  Look for the unchanged points of a transformation. In an enlargement or a rotation, the centre is unchanged. In a reflection, the mirror line is unchanged. After two reflections in non-parallel mirror lines, the intersection of these lines is unchanged, and hence this point is the centre of rotation.

## Links

A transformation question could be linked with one on ratios of lengths, areas and volumes.

# Similarity

## Definitions and results

Two figures are **similar** if they have the same shape (but not necessarily the same size). If two figures are similar, then there is a constant ratio between the corresponding sides of the two figures.

In particular, if one figure is an enlargement of another, then the two figures are similar. The constant ratio is the **scale factor**.

Two triangles are similar if one of the following conditions holds:

- they have the same angles

 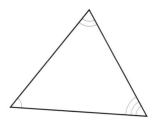

- there is a constant ratio between the sides

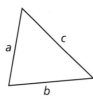 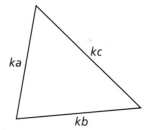

- they have an equal angle, and the sides containing the angle are in a constant ratio.

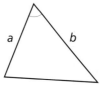 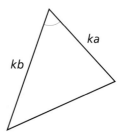

## Ratio of areas and volumes

If two objects are the same shape, with their corresponding lengths in the ratio $a:b$, then

- their areas are in the ratio $a^2:b^2$
- their volumes are in the ratio $a^3:b^3$.

*Examples* In the diagram, $AB = 6\,\text{cm}$, $AD = 4\,\text{cm}$, $AC = 9\,\text{cm}$ and $AE = 6\,\text{cm}$.

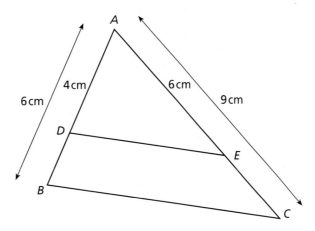

**a** Show that $DE$ is parallel to $BC$.
**b** If $BC = 8\,\text{cm}$, find $DE$.

**a** The two triangles $ABC$ and $ADE$ have $\angle A$ in common.
The ratios $AB:AD$ and $AC:AE$ are both equal to $3:2$.
Hence the triangles are similar.
Hence $\angle ADE = \angle ABC$.
Hence $DE$ is parallel to $BC$.
**b** The ratio $BC:DE$ is also equal to $3:2$.

$$DE = \tfrac{2}{3} \times 8$$

$DE = 5\tfrac{1}{3}\,\text{cm}$.

---

A solid sphere of metal has radius $4\,\text{cm}$.
**a** How many solid spheres of radius $1\,\text{cm}$ can be made from it?
**b** What is the radius of a solid sphere which has three quarters the volume of the original sphere?

**a** The ratio of the lengths is $1:4$. So the ratio of the volumes is $1^3:4^3$, i.e. $1:64$.
Each small sphere has volume $\tfrac{1}{64}$ of the original sphere.
64 small spheres can be made.
**b** Suppose the sphere has radius $r\,\text{cm}$. Then $r^3:4^3 = 3:4$.

$$r^3 = 64 \times \tfrac{3}{4} = 48$$

Take the cube root of 48.
The radius is $3.63\,\text{cm}$.

## Exercise on similarity

**1** $D$ and $E$ are points on the sides $AB$ and $AC$ respectively of $\triangle ABC$, with $DE$ parallel to $BC$. If $BC = 21\,\text{cm}$, $AB = 18\,\text{cm}$ and $AD = 15\,\text{cm}$, find $DE$.

**2** $D$ and $E$ are points on the sides $AB$ and $AC$ respectively of $\triangle ABC$. $AB = 20\,\text{cm}$, $AD = 18\,\text{cm}$, $AC = 30\,\text{cm}$ and $AE = 12\,\text{cm}$. If $DE = 15\,\text{cm}$, find $BC$.

3 In the diagram, $AB$ is parallel to $CD$. $BX = 4$ cm, $AB = 5$ cm and $CX = 10$ cm. Find $CD$.

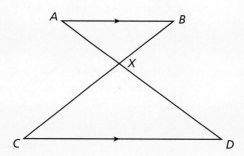

4 In question 3, the area of $\triangle ABX$ is $7 \text{ cm}^2$. What is the area of $\triangle CDX$?
5 $ABCD$ is a parallelogram, and $M$ is on $BC$ with $BM:MC = 3:2$. If $\triangle AXD$ has area $42 \text{ cm}^2$ find the area of $\triangle MXB$.

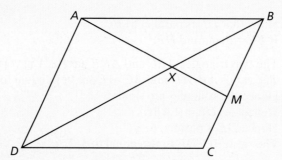

6 Two discs are cut from the same sheet of metal. The larger disc has radius 8 cm and mass 27 grams. The smaller disc has radius 6 cm. What is the mass of the smaller disc?
7 A cylindrical can of height 10 cm contains 1000 ml of liquid. A larger size is similar in shape, but has height 15 cm. Find the volume of the contents of the larger can.
8 A marble statue is 3.2 m high. A model of the statue, 10 cm high, is made from the same marble. If the model has mass 80 grams, find the mass of the original statue.
9 Two squares are cut from the same sheet of cardboard. The smaller square has side 15 cm and mass 3.6 grams. The larger square has mass 6.4 grams. What is its side?
10 Two solid spheres are made from the same metal. The larger has radius 0.4 m and mass 1600 kg. The smaller has mass 400 kg. What is the radius of the smaller sphere?

## Question spotting
The word 'similar' may not occur in the question. Watch out for a question which asks you to find lengths, but with insufficient information to use Pythagoras' theorem or trigonometry.

## Tips
When using similarity, make sure you get the sides in the correct order.
  Don't forget to square or cube the length ratio when finding the area ratio or volume ratio.

## Links
A question on similarity could be mixed with one on areas and volumes. It could also be part of a longer question on other topics of geometry.

# Three-dimensional geometry

Trigonometry and Pythagoras' theorem can be used to find distances and angles in solid objects.

## Method

Find a two-dimensional section of the solid. Then use Pythagoras' theorem or trigonometry in two dimensions.

The angle between a line and a plane is found by drawing a perpendicular from the line to the plane. The angle is $\theta$ as in the diagram.

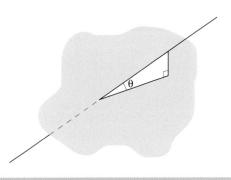

*Example*   $ABCDEFGH$ is a cuboid with $AE = 4$ cm, $AD = 5$ cm and $DC = 6$ cm. Find

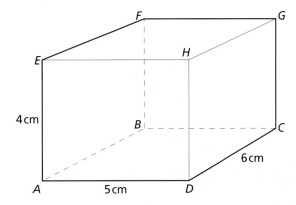

**a** $AG$
**b** the angle between $AG$ and $GD$
**c** the angle between $AG$ and the base $ABCD$.

**a** The section to use is $AEGC$. We know that $GC = 4$ cm. $AC$ is the diagonal of the base $ABCD$.

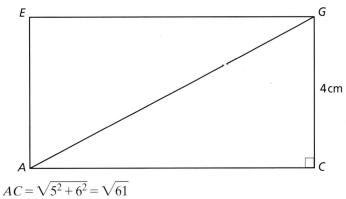

$$AC = \sqrt{5^2 + 6^2} = \sqrt{61}$$
$$AG = \sqrt{4^2 + (\sqrt{61})^2} = \sqrt{16 + 61} = \sqrt{77}$$

$AG = 8.77$ cm.

**b** We need the section $AFGD$.

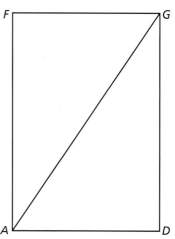

We know that $AD = 5$ cm and we have just found that $AG = \sqrt{77}$ cm.

$$\angle AGD = \sin^{-1}\left(\frac{5}{\sqrt{77}}\right)$$

The angle between $AG$ and $GD$ is $34.7°$.

**c** The perpendicular from $G$ meets $ABCD$ at $C$. So we want $\angle GAC$. Refer to the diagram for part **a**.

$$\angle GAC = \sin^{-1}\left(\frac{4}{\sqrt{77}}\right)$$

The angle between $AG$ and $ABCD$ is $27.1°$.

## Exercise on three-dimensional geometry

**1** $ABCDEFGH$ is a cuboid with $AB = 7$ cm, $AE = 8$ cm and $BC = 10$ cm. Find
  **a** $EC$
  **b** the angle between $EC$ and $CB$
  **c** the angle between $EC$ and $EFBA$.

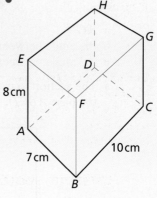

**2** $VABCD$ is a pyramid with $ABCD$ a square base of side $56$ m. $VA = VB = VC = VD = 63$ m. Find
  **a** the height of $V$ above $ABCD$
  **b** the angle between $VA$ and $VC$
  **c** the angle between $VA$ and $ABCD$.

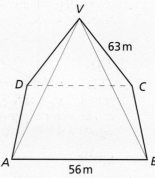

**3** $VABCD$ is a pyramid, with a rectangular base $ABCD$ for which $AB = 32\,\text{cm}$ and $AD = 45\,\text{cm}$. The vertex $V$ is $54\,\text{cm}$ above the centre $X$ of the base. Find
 **a** $VA$
 **b** the angle between $VA$ and $VB$
 **c** the angle between $VA$ and $ABCD$.

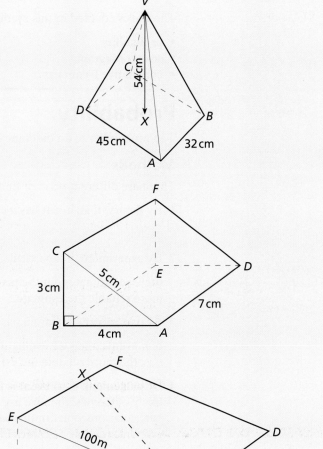

**4** $ABCDEF$ is a prism, in which $ABC$ is a right-angled triangle with sides $3\,\text{cm}$, $4\,\text{cm}$ and $5\,\text{cm}$. $AD = 7\,\text{cm}$. Find
 **a** $AF$
 **b** the angle between $AF$ and $AC$
 **c** the angle between $AF$ and $ADEB$.

**5** The diagram shows a flat hillside which slopes at $23°$ to the horizontal. Liam rides his mountain bike in a straight line up the hill, at a gradient of $17°$ to the horizontal. He starts from $A$ and ends at $X$ on the ridge $EF$. How far is $X$ from $A$?

**6** The Second Pyramid at Giza in Egypt has a square base of side $216\,\text{m}$ and a height of $144\,\text{m}$. Find.
 **a** the length of each sloping edge
 **b** the angle between a sloping edge and the horizontal
 **c** the volume of the pyramid
 **d** the total surface area of the triangular faces.

## Question spotting
You will probably be provided with a diagram of the three-dimensional object. If the cross-sections you use do not have a right-angled triangle, then you will have to use the sine or cosine rule, as in chapter 2.

## Tips
These questions often have a part **a** which leads on to part **b**. Write down the rounded answer to part **a**, but keep it more accurately when you use it for part **b**. This was done in the example above – the answer to part **a** was given to three significant figures, but for parts **b** and **c** we used the accurate value of $\sqrt{77}$.

## Links
This topic could be connected with that of volumes and areas.

# Review module 4 – Handling data

The topics covered in this module are

- **probability**
- **scatter diagrams**
- **cumulative frequency.**

## Probability

The probability of an outcome is a measure of how likely we think it is.

### Methods

There are different ways of finding probability.

1 By theory: if an event has $n$ equally likely outcomes, then the probability of each is $\frac{1}{n}$.

2 By experiment: if a particular outcome happens $r$ times out of $n$ experiments, then we estimate that the probability of the outcome is $\frac{r}{n}$. The larger $n$ is, the more reliable the estimate.

3 By past records: if, in the past, a particular outcome has occurred $r$ times out of $n$ events, then we estimate the probability of the outcome is $\frac{r}{n}$. The larger $n$ is, the more reliable the estimate.

4 By judgement: if an event is unrepeatable, and cannot be split up into equally likely outcomes, then we use our judgement to estimate the probability of a particular outcome. In this case, the result is subjective. Different people may give different probabilities for the outcome.

### Results

- If the probability of $A$ is $p$, then the probability of $A$ not happening is $1 - p$.
- Two outcomes are **mutually exclusive** if they cannot happen together. If $A$ and $B$ are mutually exclusive, then $P(A \text{ or } B) = P(A) + P(B)$.
- Two outcomes are **independent** if the result of one does not alter the probability of the other. If $A$ and $B$ are independent, then $P(A \text{ and } B) = P(A) \times P(B)$.
- Suppose an outcome has probability $p$. If the event is repeated $n$ times, the probability that the outcome occurs each time is $p^n$. The probability that it occurs at least once is $1 - (1-p)^n$.

*Examples*  A fair dice is rolled ten times. What is the probability that
**a** the dice comes up 6 all ten times
**b** the dice comes up 6 at least once.

**a** For all outcomes to be 6, take the tenth power of $\frac{1}{6}$.
P(ten 6s) = $1.65 \times 10^{-8}$.

**b** The probability that a roll doesn't give 6 is $\frac{5}{6}$. The probability that none of the ten rolls gives a 6 is $(\frac{5}{6})^{10}$. Hence the probability of at least one 6 is 1 minus this.
P(at least one 6) = 0.838.

How many times must a fair dice be rolled before we are 99% sure that the outcomes won't all be 6s?

After $n$ rolls, the probability that the outcomes are all 6s is $(\frac{1}{6})^n$. We want to ensure that this probability is less than 0.01.

$$(\tfrac{1}{6})^2 = \tfrac{1}{36} \qquad (\tfrac{1}{6})^3 = \tfrac{1}{216} < 0.01$$

We must roll the dice three times.

---

The table below gives the number of cars and the number of adults in a group of 440 households.
a  If a household is picked at random, what is the probability that it has 1 car and 2 adults?
b  If a household is picked at random, what is the probability that it has 1 car?
c  What is the probability that a house with 3 adults has 2 cars?
d  What is the mean number of cars per household?

|  | number of adults |  |  |  |  |  |
|---|---|---|---|---|---|---|
| number of cars | 1 | 2 | 3 | 4 | 5 | 6 |
| 0 | 23 | 47 | 31 | 12 | 3 | 0 |
| 1 | 42 | 75 | 66 | 9 | 6 | 1 |
| 2 | 0 | 38 | 41 | 12 | 5 | 3 |
| 3 | 0 | 0 | 12 | 9 | 3 | 2 |

a  There are 75 households with 1 car and 2 adults.
   The probability is $\frac{75}{440}$, which is $\frac{15}{88}$.
b  There are 199 households with 1 car.
   The probability is $\frac{199}{440}$.
c  There are 150 households with 3 adults. Of these, 41 have 2 cars.
   The probability is $\frac{41}{150}$.
d  To find the mean, find the total number of cars and divide by 440. The row totals give the numbers of households with 0 cars, 1 car etc. These row totals are 116, 199, 99 and 26.

$$(0\times 116 + 1\times 199 + 2\times 99 + 3\times 26) \div 440 = 1.08$$

The mean number of cars per household is 1.08.

# Exercise on probability

1  How would you find the probabilities of the following?
   a  That a randomly chosen day in August will be wet
   b  That a particular boxer will win a boxing match
   c  That a fair spinner will come down on one particular edge
   d  That a soup can, when knocked over, will land on a flat face
2  A bag contains 8 red, 5 blue and 7 yellow beads. One is drawn.
   a  What is the probability that it is red?
   b  What is the probability that it is not yellow?
   c  What is the probability that it is red or blue?
   d  What is the probability that it is red and blue?

3 A bag contains 8 red and some blue counters. When a counter is drawn, the probability that it is blue is $\frac{13}{17}$. How many blue counters are there?

4 To start a game, a five-sided spinner is spun until the edge with 1 on it is obtained.
  a Find the probability that three spins are required.
  b Find the probability that more than six spins are required.
  c Show that one is 90% certain of starting within eleven spins.

5 Kieron has probability $\frac{1}{4}$ of passing his driving test. This probability does not change, no matter how many times he takes the test.
  a Find the probability that he takes the test three times before passing.
  b Find the probability that he takes more than four tests.
  c Find how many tests he must take before he has a 50% chance of passing.

6 Refer to the example on page 301 about the numbers of adults and cars in 440 households.
  a If a household is picked at random, what is the probability that it has 4 adults and 3 cars?
  b If a household is picked at random, what is the probability that it has 2 adults?
  c What is the probability that a household with 2 cars has 4 adults?
  d What is the probability that there are more adults than cars in a household?
  e Find the mean number of adults per household.

7 A football team plays matches at its home ground, its opponent's ground or a neutral ground. The result could be a win, a draw or a loss. The table below gives the results after 84 matches.

|      | home | opponent's | neutral |
|------|------|------------|---------|
| win  | 19   | 9          | 3       |
| draw | 7    | 11         | 2       |
| loss | 13   | 16         | 4       |

  a A game is picked at random.
    i) What is the probability that it is a draw at the home ground?
    ii) What is the probability that it is a draw?
  b What is the probability that a win picked at random was on a neutral ground?
  c What is the probability that a game played at home ended in a win?

### Question spotting
The words 'probability' or 'chance' should occur in a question on this topic. In none of these questions is a tree diagram necessary. (Tree diagrams were covered in chapter 4.)

### Tips
Don't use the rules for combining outcomes with *and* or *or*, unless you know that the outcomes are independent or mutually exclusive respectively.

### Links
Questions about two-way tables could be phrased in terms of proportion rather than probability. For example, part **b** of the example on page 301 could ask: 'What is the proportion of houses with 1 car?'.

# Scatter diagrams

This topic is about the connection between two quantities.

## Definitions and methods

Suppose we have two quantities which may or may not be connected (such as height and weight). When we plot points corresponding to the two quantities, we obtain a **scatter diagram** (sometimes called a scatter graph). If the points lie roughly on a line with positive gradient, there is **positive correlation**. If they lie roughly on a line with negative gradient, there is **negative correlation**. If there neither of these cases hold, there is **zero correlation**. See the diagram below.

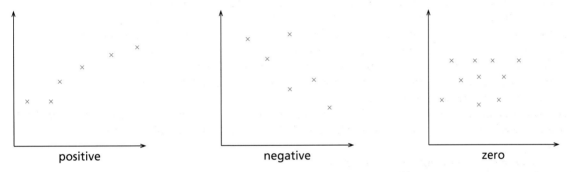

A straight line going through the points is a **line of best fit**. This line can be used to make predictions. The predictions are reliable for values in between the original points, and unreliable for values far away from the original points.

**Note.** Two quantities can be connected, even if there is zero correlation between them. In the diagram $y$ is related to $x$ by $y = x^2$, but there is zero correlation between $y$ and $x$.

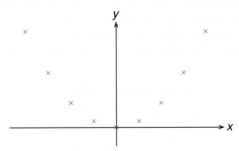

*Example*   Ten internet users were asked how long they had been using the internet, and how much time they had spent connected over the previous week. The results are below.

| length of use (years) | 1.3 | 2.8 | 0.9 | 5.3 | 6.0 | 1.7 | 4.0 | 5.1 | 3.3 | 3.8 |
|---|---|---|---|---|---|---|---|---|---|---|
| time connected (hours) | 11.7 | 9.0 | 10.6 | 5.2 | 1.5 | 8.7 | 6.5 | 8.2 | 4.1 | 3.4 |

a  Plot these points on a scatter diagram. What sort of correlation is there?
b  Draw a line of best fit through the points.
c  Brendan spent 8 hours surfing last week. How long do you think he has been using the internet?
d  Kathy has been using the internet for 7.2 years. How long do you think she spent connected last week?
e  Which of the results of **c** and **d** is more reliable, and why?

a The points are plotted on a diagram. The points are on a slope going downwards. There is negative correlation.

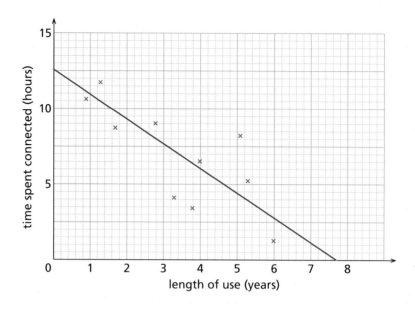

b A line of best fit is shown.
c On the line of best fit, a time of 8 hours corresponds to 2.8 years.
  Brendan has been using the internet for 2.8 years.
d On the line of best fit, using the internet for 7.2 years corresponds to a time of 0.75 hours.
  Kathy spent 0.75 hours connected.
e The result of **c** is more reliable, as it comes from a value in the middle of the other points.

**Note.** There is no single correct line of best fit. Different people will draw different lines, and get slightly different answers for **c** and **d**.

## Exercise on scatter diagrams

**1** A course is assessed by a combination of coursework and final exam. The results of eight students are below.

| coursework | 8  | 9  | 16 | 17 | 5  | 10 | 13 | 14 |
|------------|----|----|----|----|----|----|----|----|
| exam       | 42 | 40 | 68 | 78 | 39 | 50 | 72 | 60 |

a Plot these values on a scatter diagram. What sort of correlation is shown?
b Draw a line of best fit on your diagram.
c From your line, predict
  i) the exam mark of a candidate who got 12 in the coursework
  ii) the coursework mark of a candidate who got 30 in the exam.
d Which of the results in **c** is more reliable? Give reasons.

**2** In an experiment, seven batches of seeds were planted and kept at different temperatures. The temperatures and germination rates are given below.

| temperature (°C) | 5 | 10 | 15 | 20 | 25 | 30 | 35 |
|---|---|---|---|---|---|---|---|
| germination rate (%) | 24 | 31 | 38 | 45 | 51 | 75 | 72 |

  **a** Plot these values on a scatter diagram. What sort of correlation is shown?
  **b** Draw a line of best fit on your diagram.
  **c** Predict the temperature for a 60% germination rate.
  **d** Predict the germination rate at a temperature of 17 °C.

**3** Over eight successive Augusts, the total rainfall and the sales of ice cream in a resort were found. The results are below.

| rainfall (mm) | 63 | 32 | 0 | 8 | 89 | 53 | 28 | 42 |
|---|---|---|---|---|---|---|---|---|
| ice cream sales (1000s) | 58 | 92 | 106 | 110 | 50 | 48 | 77 | 70 |

  **a** Plot these values on a scatter diagram. Describe the correlation shown.
  **b** Draw a line of best fit on your diagram.
  **c** How many ice creams do you think were sold in an August when there was 65 mm of rain?

**4** Below are three situations and three scatter diagrams. Which diagrams fits which situation?
  **a** The prices of 10 cars and their ages
  **b** The ages and prices of 10 bottles of vintage wine
  **c** The maths GCSE mark and the time taken by 10 students to run 100 m

i) 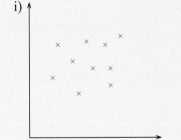   ii)    iii)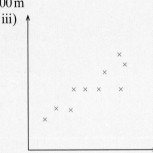

## Question spotting

One of the phrases 'scatter diagram' or 'line of best fit' is likely to occur in the question.

## Tips

The line of best fit should be a single straight line through the points. It must not zigzag from one point to the next.

The line does not have to go through any of the points. In particular, it need not go through the first and the last point. As a rough guide, ensure that there are about as many points above the line as below.

## Links

This topic is likely to occur on its own.

# Cumulative frequency

## Definitions

- The **cumulative frequency** of data at a value is the number of data less than the value.
- A graph of the cumulative frequencies against the values is a **cumulative frequency curve**.

**Note.** The cumulative frequency can also be defined as the number less than *or equal to* the value. Unless there are very few values, there is little difference between the definitions.

## Methods

The cumulative frequency curve can be used to find the median and quartiles. These are the values for which the cumulative frequency is $\frac{1}{2}$, $\frac{1}{4}$ and $\frac{3}{4}$ of the total frequency.

*Example*  The frequency table gives the amounts, £$a$, spent in a week by 100 people.
**a** Fill in the column for the cumulative frequencies.
**b** Draw a cumulative frequency curve.
**c** Estimate the number of people who spent at least £90.
**d** Find the median amount spent.
**e** Find the interquartile range of the amount spent.

| amount (£) | frequency | cumulative frequency |
|---|---|---|
| $0 \leq a < 40$ | 7 | |
| $40 \leq a < 80$ | 23 | |
| $80 \leq a < 120$ | 31 | |
| $120 \leq a < 160$ | 26 | |
| $160 \leq a < 200$ | 13 | |

**a** The cumulative frequencies are the running totals of the frequencies.

| amount (£) | frequency | cumulative frequency |
|---|---|---|
| $0 \leq a < 40$ | 7 | 7 |
| $40 \leq a < 80$ | 23 | 30 |
| $80 \leq a < 120$ | 31 | 61 |
| $120 \leq a < 160$ | 26 | 87 |
| $160 \leq a < 200$ | 13 | 100 |

**b** Plot points at the right limits (upper limits) of each interval. So plot at (40, 7), (80, 30), (120, 61), (160, 87) and (200, 100). Join the points up with a smooth curve.

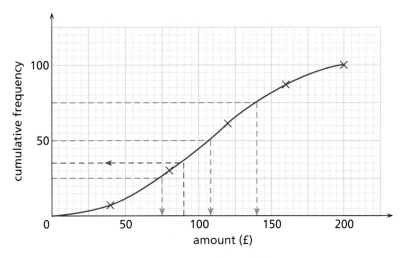

**c** A point on the graph is (90, 35). So 35 people spent less than £90. Subtract 35 from 100.
65 people spent at least £90.

**d** Half the total frequency is 50. The point on the graph with cumulative frequency 50 is (108, 50).
The median is £108.

**e** The quartiles are where the cumulative frequencies are 25 and 75. These are at 75 and 140 on the horizontal axis. Subtract 75 from 140.
The interquartile range is £65.

**Note.** The last cumulative frequency is 100, which is the total number of people. This is a check that the arithmetic is correct.
The significance of the interquartile range is that the 'middle half' of the people had a range of £65 in their spending.

## Exercise on cumulative frequency

**1** The table below gives the playing times, $t$ minutes, of 200 CDs.

| playing times | frequency |
|---|---|
| $45 \leq t < 50$ | 11 |
| $50 \leq t < 55$ | 26 |
| $55 \leq t < 60$ | 68 |
| $60 \leq t < 65$ | 53 |
| $65 \leq t < 70$ | 42 |

**a** Write out a separate column for the cumulative frequencies.
**b** Draw a cumulative frequency curve.
**c** Find the median and the quartiles.
**d** What proportion of the CDs had playing times over 67 minutes?

**2** In an investigation concerning people's ideas of time, 80 people were asked to indicate when they thought a minute had passed. The results, s seconds, are given below.

| time, s (seconds) | 30 ≤ s < 40 | 40 ≤ s < 50 | 50 ≤ s < 60 | 60 ≤ s < 70 | 70 ≤ s < 80 |
|---|---|---|---|---|---|
| frequency | 8 | 15 | 31 | 17 | 9 |

    **a** Write out a separate row for the cumulative frequencies.
    **b** Draw a cumulative frequency curve.
    **c** What proportion of people were within 5 seconds of the correct time?
    **d** Find the median and the interquartile range of the times.

**3** 100 batteries were tested by being put in a toy. The time they lasted is given in the table below.

| time, h (hours) | 2 ≤ h < 3 | 3 ≤ h < 4 | 4 ≤ h < 5 | 5 ≤ h < 6 |
|---|---|---|---|---|
| frequency | 16 | 38 | 31 | 15 |

    **a** Write out a separate row for the cumulative frequencies.
    **b** Draw a cumulative frequency curve.
    **c** Estimate the number of batteries which lasted longer than $3\frac{1}{2}$ hours.
    **d** Find the median, quartiles and interquartile range.

**4** A group of 60 athletes ran 400 m twice, before and after a period of training. The cumulative frequency curves show the results.

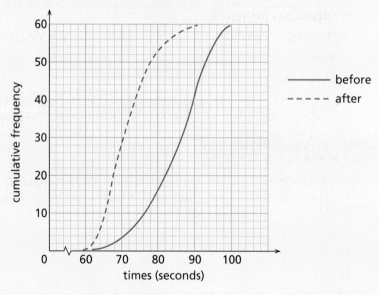

    **a** For each curve, find the median and the interquartile range. Comment on the difference.
    **b** Kieron took 78 seconds to run the first race. How long might he have taken to run the second race?

5 1000 candidates enter an exam which includes coursework and a final exam. The diagram shows cumulative frequency curves for both components.

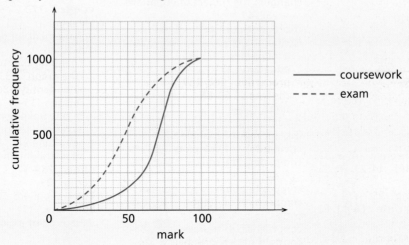

a For each curve, find the median and the interquartile range. Comment on the difference.
b Mary got 83% for coursework, but missed the exam because of illness. What would be a fair mark to give her for the exam?

### Question spotting

The phrase 'cumulative frequency' should appear on any question from this topic.

### Tips

Make sure you plot the points at the *right limits* of the intervals, not in the middles. When finding the median of 1000 numbers, the result is not 500, it is the 500th number.

### Links

A cumulative frequency question might come all by itself. It might come in connection with a histogram question for the same data, or a question asking you to estimate the mean.

# Index

## A

accuracy
  appropriate 237–9
  limits 234–5
Achilles 174
addition
  fractions 4
  vectors 194
algebraic expressions
  expansion 1, 7
  factorising 7
  fractions 4-5, 7
alternate segment theorem 76–8, 82
angle-side-angle (ASA) 14, 23, 28, 88
angles
  of 60°, constructing 95
  at the centre of a circle 68–71, 82
  bisecting 95, 272–3
  of depression 9
  of elevation 8
  exterior 82
  external 74
  finding 157–60
  negative 160–1
  opposite 74, 82
  over 360° 162
  reflex 68, 71
  results 266–7
  on the same arc 71–3, 82
  in a semicircle 64, 71
approximation 236–7
arc
  angles on 71–3, 82
area 282–8
  ratio 294–295
arithmetic, of vectors 194–198
arithmetic sequences see sequences
ASA see angle-side-angle
asymptotes 153, 166
average, moving 188–9, 190, 192

## B

Babylonian mathematics 30, 46
back-to-back diagrams 111–12
bearings 9, 168
Bernoulli, Jacob 62
bias 186, 190
birthday problem 61
borrowing 223
box plots (box-and-whisker diagrams) 105–109, 113
Broadcasters' Audience Research Board (BARB) 182

## C

calculation, accurate method 129
calculator
  degree mode 9
  probability calculations 183

centroid, of triangle 131
changing the subject of equations 252–4
chord, bisector 63
circles
  coordinates 127–9
  equation 131
  properties 282, 283
  theorems 63–85
  unit 154–6
circumcentre, of triangle 131
circumcircle, of triangle 274
comparison, of box plots 107–9
completing the square 34–5, 38–9, 44
composite numbers see numbers
compression of graphs 148, 163
cone, properties 284
congruence, of triangles 23, 88–9
constant of proportionality 242
construction
  locus 272, 274–5
  review 272–3
  triangles 8, 31
coordinate geometry, and vectors 198–201
coordinates 117–31
correlation, positive, negative or zero 303
cos see cosine
cosine
  for all angles 152, 166
  definition 154
  graph 153, 166
cosine rule 19–22
  finding unknown angles 21–2
  finding unknown sides 20–1
  side-angle-side (SAS) 28
  side-side-side (SSS) 28
  in three dimensions 25–9
  vector direction 197
counterexample 86–7
cuboid, properties 284
curve, cumulative frequency 105, 306
cylinder, properties 284

## D

data, five indicators 105
decimals, recurring and terminating 209–10, 216
denominator, of a fraction 230
density 44, 284
difference, common 170, 179
dimensions 289–90
disproof 87
distance between points 123–5
division, of fractions 4, 230
dot product 206
drawing, with or without replacement 57–9, 60

## E

eclipses 172
elimination 125
enlargement, of graphs 290
equations
  changing the subject 252–4
  of a graph 133
  and identity 86
  linear 125–7
  in one unknown 245–7
  quadratic see quadratic equations
  simultaneous 125–7, 248–50
  of a straight line 117
errors 234–5
Euclid 86
Euler, Leonhard 85, 131
Euler line 131
expansion of algebraic expressions 1, 7
exponential (geometric) sequences see sequences
expressions with four terms, factorising 5–6

## F

factorisation
  algebraic expressions 2–3, 7
  checking by expansion 2, 6
  expressions with four terms 5–6
  quadratics 2–3
  solving quadratic equations 32–3
factors
  highest common (HCF) 2, 232
  and multiples, review 232–3
  prime 232
Fermat, Pierre de 180
Fibonacci 180
foreign currency 222
fractions
  addition 4, 230
  algebraic 4–5, 7
  division 4, 230
  improper 230
  multiplication 4, 230
  as rational numbers 208
  review 230–1
  subtraction 4, 230
frequency
  cumulative 105, 306–9
  density 99, 113
  relative 182
function, objective 258

## G

Galileo 177
geometric sequences see sequences, exponential
geometry
  coordinate 117–31, 198–201

three dimensional 297–9
vectors 194–8, 201–4
golden ration 47
gradient 117, 120, 129
  product 120, 124
graphs
  compressing 148
  cosine 153, 167
  drawing/plotting 37–8, 260–5
  enlargement 290
  equation 133
  linear (straight line) 117, 170
  maximum and minimum values 36–7
  reflections 144–6, 148, 290
  rotation 290
  sine 152, 155–6, 166
  solution of quadratic equations 33
  stretching 140–144, 148
  tan 153
  transformations 132–151, 163–166
  translations 132–139, 148, 290
  trigonometric 163–6

## H
Hero of Alexandria 30
highest common factor (HCF) see factors
hire purchase 221–222, 226
histograms 99–104, 113
hypotenuse 8, 277

## I
identity, and equation 86
incircle, of triangle 274
income tax 224–6
inequalities 254–7
intercept 117
interest, simple and compound 173, 220–1, 226
irrational numbers 210

## K
kite
  properties 267
  tangent 63

## L
law of large numbers (law of averages) 62
length 282–8
linear (arithmetic) sequences see sequences
linear equations see equations
linear programming see programming
linear sequences see sequences

lines
  of best fit 303
  definition 272
  parallel 120–2, 129
  perpendicular 120–2
  segment 272
  straight, equation 117
locus, constructing 272, 274–5

## M
maximum 36–7, 44
median 105, 203
Megalithic sites 30
minimum 36–7, 44
money 219–28
mortgages 223, 228
moving average see average
multiple
  least common (LCM) 232
  see also factors and multiples
multiplication
  fractions 4, 230
  vectors 195–6, 201, 206
musical notes, sine wave 152, 163, 168

## N
numbers
  composite 232
  Fibonacci 180
  irrational 210–212, 216
  mixed 230
  prime 180, 232
  random 183–4
  rational 208–10, 216
  standard form 239–42
  triangular 169
numerator, of a fraction 230

## O
objective function 258
ogive 105
opinion polls 182, 183
order
  of operations 288, 293
  of symmetry 291
orthocentre, of triangle 130
outcomes
  independent 48, 60, 300
  mutually exclusive 48, 60, 300
overtime 219, 226

## P
palindromic number 96
parallel see lines, parallel
parallelepiped 207
parallelogram
  of midpoints 201, 204
  properties 267, 283
Pascal, Blaise 218

perpendicular
  bisector 95, 272
  constructing 94–5, 272
  see also lines, perpendicular
Plato 86
points, in moving averages 188, 190
population, sampling 183, 190
position, vector 198–200, 201, 204
powers 3, 7
primes 180, 232
prism, properties 284
probability 48–62, 300–2
  calculators 183
  conditional 49, 60
  experimental 62, 182
  spreadsheet 62
  theory 182
profit and loss 220
programming, linear 258–9
proof
  algebraic 96–8
  by contradiction 211
  and counterexample 86–7
  geometric 88–95
  and verification 86
proportion 242–4
  direct 242
  inverse 242
proportionality, constant 242
public lending right 219
pyramid, properties 284
Pythagoras' theorem
  and coordinates 123
  history and prehistory 30
  review 277–82
  surds 214–15
  three dimensional 297–9
  and vector magnitude 197

## Q
quadrants 155, 166
quadratic equations
  definition 2
  practical uses 42–4
  simultaneous 125–7
  solution
    by completing square 38–9
    by factorising 32–3
    by formula 40–4
    graphical 33
    methods 41
  spreadsheet use 47
quadratic expressions 2, 32–47
  factorising 2–3
quadratic formula 40–1, 44
quadratic sequences see sequences
quadrilaterals
  cyclic 74–6, 82
  results 267–8
quartiles 105

# 312 INDEX

## R
random number generator  183
rate of exchange  222, 226
ratio
  areas and volumes  294–5
  common  173, 179
rational numbers  208
rectangle, properties  268, 282, 283
reflection, graphs  144–6, 148, 290
rhombus  268
Roosevelt, Franklin D.  186
roots *see* surds
rotation, graphs  290

## S
salary *see* money
samples
  bias  186, 190
  from a population  183–6, 190
  random  183–4, 190
  size  182
  statistics  182–92
  stratified  184–5, 190
Saros period  172
SAS *see* side-angle-side
scalar
  definition  193
  product  195–6, 206
scale factor  294
scatter diagrams  303–5
sector, properties  283
segment, properties  283
sequences
  exponential (geometric)  173–175, 179
  Fibonacci  180
  linear (arithmetic)  170–172, 179
  quadratic  175–6, 179
  rule or formula  169
  spreadsheet use  178, 181
sexagesimal fractions  46
side-angle-side (SAS)  19, 23, 28, 88
side-side-angle (SSA)  23, 28, 88
side-side-side (SSS)  23, 28, 88
similarity  294–6
simplification  3, 4, 7
simultaneous equations *see* equations; quadratic equations
sin (trigonometric function) *see* sine
sine
  for all angles  152, 166
  definition  154
  graph (curve)  152, 155–6, 166, 168
sine rule  14–18
  angle-side-angle (ASA)  28
  finding unknown angles  17–18
  finding unknown sides  15–17
  side-side-angle (SSA)  28
  three dimensions  25–9
  vector direction  197

slope *see* gradient
SOHCAHTOA (mnemonic), trigonometric functions  152
sphere, properties  283, 284
spreadsheet
  mortgages  228
  moving average  192
  probability experiments  62
  quadratic equations  47
  random number generator  183
  sequences  178, 181
square
  completing  34–5, 38–9, 44
  properties  268
  roots *see* surds
SSA *see* side-side-angle
SSS *see* side-side-side
standard (index) form, numbers  239–242
statistics
  diagrams  98–116
  sampling  182–92
stem-and-leaf diagrams  110–12, 113
straight line *see* lines
stretching, graphs  140–4, 148, 163
substitution  125, 129
subtraction
  fractions  4, 230
  vectors  194
surds  212–16
symmetry  291

## T
tan *see* tangent (trigonometric function)
tangent (to a circle)
  kite  63
  theorems  63
tangent (trigonometric function)
  for all angles  152
  graph  153
taxes  224–6
three dimensions
  geometry  297–9
  sine and cosine rules  25–9
  translations  207
time series  187–9, 190
transformations
  combined  146–8, 290
  graphs  132–51
  inverse  290
  review  290–3
  trigonometric graphs  163–166
translations
  combined  139
  graphs  132–139, 148, 290
  three dimensions  207
  vectors  193
transposition  252–254
trapezium  267, 283, 288

tree diagrams  51–56, 60
trend  187, 190
trial and improvement  250–1
triangles
  adjacent side  8
  area (Hero's formula)  30
  centroid  131
  circumcentre  131
  circumcircle  274
  congruence  23, 88–9
  construction  8, 31
  Euler centre  85
  finding unknown angles  17–18, 21–2
  finding unknown sides  15–17, 20–1
  hypoteneuse  8
  incircle  274
  opposite side  8
  orthocentre  130
  properties  283
  right-angled  8–14, 277
    *see also* Pythagoras' theorem
  solving  23–5
trigonometric functions  8, 152
  SOHCAHTOA (mnemonic)  152
  *see also* cosine; sine; tangent
trigonometry
  for all angles  152–68
  for right-angled triangles  8–14
  *see also* graphs, trigonometric
Tukey, John  110

## V
value added tax (VAT)  224, 226
vectors  193–207
  addition  194
  arithmetic  194–8
  and coordinate geometry  198–201
  definition  193
  direction  193, 197, 204
  geometry  194–8, 201–4
  magnitude  193, 197–8, 204
  multiplication  195–6, 206
  notation  194
  parallel  201
  position  198–200, 201, 204
  scalar multiplication  201
  subtraction  194
  as translations  193
verification  86
volume  282–8
  ratio  294–5

## W
wages *see* money
word problems  247, 250